Locally Flat Embeddings of 3-Manifolds in S^4

The study of smooth embeddings of 3-manifolds in 4-space has been hampered by difficulties with the simplest case, that of homology spheres. This book presents some advantages of working with locally flat embeddings. The first two chapters outline the tools used and give general results on embeddings of 3-manifolds in S^4. The next two chapters consider which Seifert manifolds may embed, with criteria in terms of Seifert data. After summarizing results on those Seifert manifolds that embed smoothly, the following chapters determine which 3-manifolds with virtually solvable fundamental groups embed. The final three chapters study the complementary regions. When these have good fundamental groups, topological surgery may be used to find homeomorphisms. Figures throughout help illustrate links representing embeddings and open questions are further discussed in the appendices, making this a valuable resource for graduate students and research workers in geometric topology.

AUSTRALIAN MATHEMATICAL SOCIETY LECTURE SERIES

All the titles listed below can be obtained from good booksellers or from Cambridge University Press. For a complete series listing visit www.cambridge.org/mathematics.

9. Algebraic Groups and Lie Groups, G. I. LEHRER (ed.)
10. Modelling with Differential and Difference Equations, G. FULFORD, P. FORRESTER & A. JONES
11. Geometric Analysis and Lie Theory in Mathematics and Physics, A. L. CAREY & M. K. MURRAY (eds.)
12. Foundations of Convex Geometry, W. A. COPPEL
13. Introduction to the Analysis of Normed Linear Spaces, J. R. GILES
14. Integral: An Easy Approach after Kurzweil and Henstock, L. P. YEE & R. VYBORNY
15. Geometric Approaches to Differential Equations, P. J. VASSILIOU & I. G. LISLE (eds.)
16. Industrial Mathematics, G. R. FULFORD & P. BROADBRIDGE
17. A Course in Modern Analysis and its Applications, G. COHEN
18. Chaos: A Mathematical Introduction, J. BANKS, V. DRAGAN & A. JONES
19. Quantum Groups, R. STREET
20. Unitary Reflection Groups, G. I. LEHRER & D. E. TAYLOR
21. Lectures on Real Analysis, F. LÁRUSSON
22. Representations of Lie Algebras, A. HENDERSON
23. Neverending Fractions, J. BORWEIN et al.
24. Wavelets: A Student Guide, P. NICKOLAS
25. Classical Groups, Derangements and Primes, T. BURNESS & M. GIUDICI
26. Notes on Counting: An Introduction to Enumerative Combinatorics, P. J. CAMERON
27. Orthogonal Polynomials and Painlevé Equations, W. VAN ASSCHE
28. Many Variations of Mahler Measures, F. BRUNAULT & W. ZUDILIN

Australian Mathematical Society Lecture Series: 29

Locally Flat Embeddings of 3-Manifolds in S^4

JONATHAN HILLMAN
University of Sydney

CAMBRIDGE
UNIVERSITY PRESS

CAMBRIDGE
UNIVERSITY PRESS

Shaftesbury Road, Cambridge CB2 8EA, United Kingdom

One Liberty Plaza, 20th Floor, New York, NY 10006, USA

477 Williamstown Road, Port Melbourne, VIC 3207, Australia

314321, 3rd Floor, Plot 3, Splendor Forum, Jasola District Centre, New Delhi 110025, India

Cambridge University Press is part of Cambridge University Press & Assessment,
a department of the University of Cambridge.

We share the University's mission to contribute to society through the pursuit of
education, learning and research at the highest international levels of excellence.

www.cambridge.org
Information on this title: www.cambridge.org/9781009715386
DOI: 10.1017/9781009715355

When citing this work, please include a reference to the DOI 10.1017/9781009715355

First published 2026

A catalogue record for this publication is available from the British Library

A Cataloging-in-Publication data record for this book is available from the Library of Congress

ISBN 978-1-009-71538-6 Paperback

Contents

Preface

Every closed 3-manifold may be smoothly embedded in S^5. The question of which 3-manifolds embed in S^4 depends markedly on the interpretation of "embedding". Freedman showed that every integral homology 3-sphere bounds a contractible topological 4-manifold, and so embeds as a topologically locally flat submanifold of S^4. This is strikingly simpler than in the case of smooth (or piece-wise linear locally flat) embeddings, where the corresponding question remains open, and seems quite delicate. The exemplary uniqueness result is the Schoenflies Theorem, which shows that all locally flat embeddings of S^{n-1} in S^n are topologically equivalent. In every dimension $n \neq 4$ the smooth analogue also holds. There is a remarkable contrast between the simplicity of the proofs in the topological case and the as-yet-unresolved status of the 4-dimensional smooth Schoenflies Conjecture.

We have chosen, therefore, to focus on topologically locally flat embeddings, rather than smooth embeddings. Many of the examples in this book are constructed by ambient surgery on "bipartite links" in the equatorial S^3, and so are smooth, but most of the algebraic constraints on embeddings that we use also obstruct locally flat embeddings into homology 4-spheres.

It seems unlikely that there will ever be a reduction of the question of which closed orientable 3-manifolds embed in S^4 to the determination of familiar invariants. However, if we restrict the scope of the question to 3-manifolds with virtually solvable fundamental groups or which are Seifert fibred, then much is known. Manifolds of the first type have one of the geometries $\mathbb{S}^3$, $\mathbb{S}^2 \times \mathbb{E}^1$, $\mathbb{E}^3$, $\mathbb{N}il^3$ and $\mathbb{S}ol^3$. Just 13 manifolds of this type embed, as was shown in joint work with Crisp. Seifert fibred manifolds either have one of the first four of these geometries or are $\mathbb{H}^2 \times \mathbb{E}^1$- or $\widetilde{\mathbb{SL}}$-manifolds. The strong results on smooth embeddings of Seifert fibred 3-manifolds due to Donald, Issa and McCoy suggest plausible outcomes for the Seifert fibred cases. These results also suggest that the study of embeddings of graph manifolds might prove rewarding.

The main themes of this book are the concentration on well-understood classes of 3-manifolds, the study of complementary regions and the use of simple invariants to show that most 3-manifolds which embed in S^4 do so in many distinct ways. We use 4-dimensional topological surgery at various points, but we make no use of gauge-theoretic ideas.

The first two chapters are introductory. In the preliminary Chapter 1 we give the notation and terminology from algebra and topology that we shall use. The key invariants are the Euler characteristic and fundamental groups of the complementary regions. We also describe the surgery exact sequence, but treat this material as a "black box" to be used without further detail. In Chapter 2 we summarize the basic facts about these invariants, which go back to the first paper on the subject (by Hantzsche in 1938). We also describe the construction of embeddings by 0-framed surgery on bipartedly slice links. This construction was introduced by Gilmer and Livingston, and was used by Lickorish and Quinn (independently) to realize pairs of groups with balanced presentations and isomorphic abelianization. We then define the notion of 2-knot surgery (first used by Livingston), by means of which it may be shown that most 3-manifolds which embed do so in infinitely many ways.

In Chapter 3 we consider 3-manifolds which are Seifert fibred over orientable base orbifolds, and apply the G-Index Theorem (for finite cyclic G) to show that if M is an $\mathbb{H}^2 \times \mathbb{E}^1$-manifold and the base orbifold has genus 0 and all cone points of odd order then M embeds in S^4 if and only if the Seifert data is skew-symmetric. We state here some of the results of Donald, Issa and McCoy. In Chapter 4 we consider 3-manifolds which are Seifert fibred over non-orientable bases and give strong bounds for the Euler numbers of manifolds with given base orbifold which can embed. (The argument here also uses the G-Index Theorem, albeit in a different manner, and for $G = \mathbb{Z}/2\mathbb{Z}$ only. Apart from this, the first four chapters use only basic algebraic topology and elementary 3-manifold topology.) In Chapter 5 we give a brief outline of work on smooth embeddings, with little by way of proofs.

In the subsequent chapters we use surgery, and for this we must restrict the groups arising as fundamental groups of 4-manifolds. Under our present understanding of the Disc Embedding Theorem, these groups should be in the class of groups generated from groups with sub-exponential growth by increasing unions and extensions. This class includes all virtually solvable groups. All 3-manifold groups in this class are in fact virtually solvable, and in Chapter 6 we determine the 3-manifolds with such groups which embed in S^4. We then show that if M is a 3-manifold such that $\pi_1(M)$ is an extension of a torsion-free solvable group G by a perfect normal subgroup then M embeds if and only if $G \cong \pi_1(P)$ for some 3-manifold P which embeds.

In the remaining three chapters we consider the complementary regions in some detail. If $M = \#^r(S^2 \times S^1)$ then M has different embeddings with complementary regions having Euler characteristics of all possible values allowed by the elementary considerations of Chapter 2. We use Massey products to show that this is not the case for M a Seifert fibred 3-manifold with $\beta_1(M) > 1$. In Chapters 8 and 9 we assume that the complementary regions have abelian or nilpotent fundamental groups (respectively). The possible groups are known in the abelian case and severely restricted in the nilpotent case. If the group is also torsion-free then there is scope for identifying the homotopy types of the complementary regions and applying surgery.

There are three appendices. In Appendix A we analyze the linking pairings of orientable 3-manifolds which are Seifert fibred over orientable base orbifolds. In

Appendix B we give some results on nilpotent groups with balanced presentations, and Appendix C is a short list of questions about embeddings of 3-manifolds in S^4.

The book is based on a number of my papers and one paper written jointly with John Crisp. These have only been cited when the presentation here is different from the earlier version. Allowing for consolidation of common material into the earlier chapters and some reorganization, Chapter 3 corresponds to [**Hi09**], Chapter 4 and the first half of Chapter 6 to [**CH98**], the second half of Chapter 6 to [**Hi96**], and the remaining chapters to [**Hi17**], [**Hi20a**], and [**Hi20b**], respectively. Appendix A is based on [**Hi11**] and Appendix B extends [**Hi22**].

I would like to thank John for the collaboration which is recorded in Chapter 4 and the first part of Chapter 6. In particular he saw how to extend unpublished work of mine on S^1-bundles to the Seifert fibred case, and then worked out how to handle the $\mathbb{S}ol^3$-manifolds which do not fibre over S^1. I would also like to thank Ray Lickorish, and Ben Burton and Ryan Budney for permission to include their proofs of Theorems 2.8 and 3.10 and Andy Putnam for allowing me to make use of his account of the proofs of the Generalised Schoenflies Theorem. Finally, I would like to thank the referees for their suggestions for improving the exposition of the material in this book.

CHAPTER 1

Preliminaries

This chapter is intended as a compendium of notation and terminology for the algebra and topology used in the subsequent chapters. Our general approach shall be to try to prove all assertions that are specifically about our topic, but to cite standard references for other supporting material. In particular, we invoke the G-Signature Theorem in the proofs of two key results (Theorems 3.9 and 4.4), and we use 4-dimensional topological surgery to provide homeomorphisms (in the later chapters). As these results and techniques may not be familiar to all 3-manifold topologists, we give condensed accounts in Sections 1.8 and 1.9.

1.1. Groups

Let G be a group. Our commutator convention is that if $g, h \in G$ then $[g,h] = ghg^{-1}h^{-1}$. We shall let ζG, $G' = [G,G]$ and $G^{ab} = G/G'$ denote the centre, commutator subgroup and abelianization of G, respectively. The lower central series is defined by $\gamma_1 G = G$ and $\gamma_{n+1}G = [G, \gamma_n G]$ for all $n \geqslant 1$. Similarly, the rational lower central series is given by letting $\gamma_1^{\mathbb{Q}} G = G$ and $\gamma_{k+1}^{\mathbb{Q}} G$ be the preimage in G of the torsion subgroup of $G/[G, \gamma_k^{\mathbb{Q}} G]$. Then $G/\gamma_k^{\mathbb{Q}} G$ is a torsion-free nilpotent group, and $\{\gamma_k^{\mathbb{Q}} G\}_{k \geqslant 1}$ is the most rapidly descending central series of subgroups of G with this property.

We may also let $G'' = [G', G']$ and $I(G) = \gamma_{[2]}^{\mathbb{Q}} G$ denote the second derived subgroup and isolator subgroup of G, respectively. (Thus $G/I(G)$ is the maximal torsion-free abelian quotient of G.) Each of the subgroups defined thus far is characteristic in G, and so is normal in any overgroup H containing G as a normal subgroup.

A group G is *restrained* if it has no non-cyclic free subgroup. It is *virtually solvable* if it has a normal subgroup of finite index which is solvable. The *Hirsch length* $h(G)$ of such a group G is the sum of the ranks of the abelian sections of a composition series for G. Amenable groups and groups which are "good" in the sense of [**FQ**] are restrained, but the most important examples for us are virtually solvable, and in fact abelian or nilpotent.

Let $F(r)$ be the free group of rank r. Let $D_\infty = \mathbb{Z}/2\mathbb{Z} * \mathbb{Z}/2\mathbb{Z}$ be the infinite dihedral group. Let $BS(1,m)$ be the Baumslag-Solitar group with presentation $\langle t, a \mid tat^{-1} = a^m \rangle$, for $m \in \mathbb{Z} \setminus \{0\}$. Then $BS(1,1) \cong \mathbb{Z}^2$, while $BS(1,-1) \cong \pi_1 Kb$ is the Klein bottle group.

Our principal reference for group theory is [**Rob**].

1.2. Homological Group Theory

The Universal Coefficient Theorems give exact sequences

$$0 \to F \otimes H_2(G;\mathbb{Z}) \to H_2(G;F) \to Tor(F, G^{ab}) \to 0$$

$$\text{and} \qquad 0 \to Ext(G^{ab}, F) \to H^2(G;F) \to Hom(G^{ab}, F) \to 0,$$

for any group G and field F, since $H_1(G;\mathbb{Z}) = G^{ab}$.

A finite presentation for a group is *balanced* if it has the same number of relations as generators. The group then has deficiency $\geqslant 0$. (We may always add trivial relators to a presentation with positive deficiency to get one which is balanced.) A finitely generated group G is *homologically balanced* if $\beta_2(G;R) \leqslant \beta_1(G;R)$, for any coefficient ring R.

If G is finitely presentable then $H_i(G;R)$ is finitely generated for $i \leqslant 2$ and all simple coefficients R. It follows easily that $\beta_i(G;\mathbb{Q}) = \beta_i(G;\mathbb{F}_p)$, for $i \leqslant 2$ and almost all primes p. Hence $\beta_2(G;\mathbb{Q}) = \beta_1(G;\mathbb{Q})$ if and only if $\beta_2(G;\mathbb{F}_p) = \beta_1(G;\mathbb{F}_p)$, for almost all primes p. If A is a finitely generated abelian group and $F = \mathbb{F}_p$ then $Tor(\mathbb{F}_p, A) \cong {}_pA = \mathrm{Ker}(p.id_A)$. If A is finite then A/pA and $\mathrm{Ker}(p.id_A)$ have the same dimension. Hence if G is finite then $\beta_2(G;\mathbb{F}_p) \geqslant \beta_1(G;\mathbb{F}_p)$, for any prime p, and G is homologically balanced if and only if $H_2(G;\mathbb{Z}) = 0$.

If G has a balanced presentation then it is homologically balanced, and if G is also finite then it must have trivial multiplicator: $H_2(G;\mathbb{Z}) = 0$. (These assertions follow most easily from consideration of the homology of the 2-complex associated to a balanced presentation for the group.) In general there may be a gap between homological necessary conditions and combinatorial sufficient conditions.

Finitely presentable restrained groups have deficiency $\leqslant 1$, and if G is restrained and $def(G) = 1$ then G is an ascending HNN extension. In the latter case the first L^2-Betti number $\beta_1^{(2)}(G) = 0$, and $c.d.G \leqslant 2$. These assertions follow from [**FMGK**, Theorem 2.5]. The argument is homological, and so it suffices that the augmentation ideal in $\mathbb{Z}[G]$ should have a presentation of deficiency 1 as a $\mathbb{Z}[G]$-module.

If A is abelian then $H_2(A;\mathbb{Z}) = A \wedge A$. A finite abelian group A is homologically balanced if and only if it is cyclic, for if A/pA is not cyclic for some prime p then $(A/pA) \wedge (A/pA) \neq 0$, and so $\beta_2(A;\mathbb{F}_p) > \beta_1(A;\mathbb{F}_p)$, by the Universal Coefficient exact sequences of §1. Finite cyclic groups clearly have balanced presentations.

A PD_3^+-group is an orientable Poincaré duality group of dimension 3.

Our principal references for cohomological group theory are [**Bie**] and [**Bro**].

1.3. Commutative Algebra

Let $\Lambda = \mathbb{Z}[\mathbb{Z}] = \mathbb{Z}[t, t^{-1}]$ and $R\Lambda = R \otimes_{\mathbb{Z}} \Lambda$, for any coefficient ring R. If F is a field then $F\Lambda$ is a PID.

If R is a PID and L is a finitely generated R-torsion module then L has a square presentation matrix, by the Structure Theorem for modules over PIDs. The *order* of L is the principal ideal generated by the determinant of any such presentation matrix for L. We shall let $\Delta_0(L)$ denote a generator of the order of L. (This is well-defined, up to multiplication by units of R.) If

$$0 \to A \to C \to B \to 0$$

is a short exact sequence of such R-modules then $\Delta_0(C) = \Delta_0(A)\Delta_0(B)$ (up to units). If $f : A \to B$ is a homomorphism then $\Delta_0(\mathrm{Im}(f)) = \Delta_0(\mathrm{Im}(\mathrm{Ext}_R(f)))$, where $\mathrm{Ext}_R(f) : \mathrm{Ext}_R(B, R) \to \mathrm{Ext}_R(A, R)$ is the induced homomorphism.

1.4. Bilinear Pairings

A *linking pairing* on a finite abelian group N is a symmetric bilinear function

$$\ell : N \times N \to \mathbb{Q}/\mathbb{Z}$$

which is non-singular in the sense that $\mathrm{ad}(\ell) : n \mapsto \ell(-, n)$ defines an isomorphism from N to $Hom(N, \mathbb{Q}/\mathbb{Z})$. If L is a subgroup of N then $\mathrm{ad}(\ell)$ induces an isomorphism $L^\perp = \{t \in N \mid \ell(t, l) = 0\ \forall l \in L\} \cong Hom(N/L, \mathbb{Q}/\mathbb{Z})$, which is non-canonically isomorphic to N/L. It is *neutral* if there is a subgroup P with $P = P^\perp$, *split* if also P is a direct summand of N and *hyperbolic* if N is the direct sum of two such subgroups. If ℓ is split then N/P is (non-canonically) isomorphic to P, and so N is a direct double. A linking pairing ℓ is *even* if $2^{k-1}\ell(x, x) \in \mathbb{Z}$ for all $x \in N$ such that $2^k x = 0$. Hyperbolic pairings are even. We shall say that ℓ is *odd* if it is not even.

Every linking pairing splits uniquely as the orthogonal sum (over primes p) of its restrictions to the p-primary subgroups of N. If $w = \frac{p}{q} \in \mathbb{Q}^\times$ (where $(p, q) = 1$) let ℓ_w be the pairing on $\mathbb{Z}/q\mathbb{Z}$ given by $\ell_w(m, n) = [mnw] \in \mathbb{Q}/\mathbb{Z}$. Then $\ell_w \cong \ell_{w'}$ if and only if $w' = n^2 w$ for some integer n with $(n, q) = 1$. When $q = p^k$ is a power of an odd prime there are just two isomorphism classes of such pairings. Thus every linking pairing on an abelian group of odd order is an orthogonal sum of pairings on cyclic groups. However, if $q = 2^k$ then $\ell_w \cong \ell_{w'}$ if and only if $2^k w' \equiv 2^k w$ *mod* $(2^k, 8)$. In this case there are also indecomposable pairings E_0^k and E_1^k on the groups $(\mathbb{Z}/2^k\mathbb{Z})^2$, with matrices $\begin{pmatrix} 0 & 2^{-k} \\ 2^{-k} & 0 \end{pmatrix}$, for $k \geqslant 1$, and $\begin{pmatrix} 2^{1-k} & 2^{-k} \\ 2^{-k} & 2^{1-k} \end{pmatrix}$, for $k \geqslant 2$, respectively. The set of all such pairings is a semigroup with respect to orthogonal direct sum, and its structure has been completely determined [**KK80, Wa64**].

It is often convenient to study linking pairings via matrices. Let $N \cong (\mathbb{Z}/p^k\mathbb{Z})^\rho$, with basis $e_1, \dots, e_\rho$, and let ℓ be a linking pairing on N. Let L be the $\rho \times \rho$ matrix with (i, j) entry $p^k \ell(e_i, e_j)$, considered as an element of $\mathbb{Z}/p^k\mathbb{Z}$. Then $L \in \mathrm{GL}(\rho, \mathbb{Z}/p^k\mathbb{Z})$, since ℓ is non-singular. The *rank* of ℓ is $rk(\ell) = \dim_{\mathbb{F}_p} N/pN = \rho$. If p is odd then a linking pairing ℓ on a free $\mathbb{Z}/p^k\mathbb{Z}$-module N is determined up to isomorphism by $rk(\ell)$ and the image $d(\ell)$ of $\det(L)$ in $\mathbb{F}_p^\times/(\mathbb{F}_p^\times)^2 = \mathbb{Z}/2\mathbb{Z}$. (This is independent of the choice of basis for N.) In particular, ℓ is hyperbolic if and only if $\rho = rk(\ell)$ is even and $d(\ell) = [(-1)^{\frac{\rho}{2}}]$. If $p = 2$ and $k \geqslant 3$ then ℓ is determined by the image of L in $GL(\rho, \mathbb{Z}/8\mathbb{Z})$; if moreover ℓ is even and $k \geqslant 2$ then ρ is even and ℓ is determined by the image of L in $GL(\rho, \mathbb{Z}/4\mathbb{Z})$ [**De05, KK80, Wa64**].

1.5. Algebraic Topology

If W is a cell-complex its universal cover $\widetilde{W}$ has an induced cellular structure. Let $\Gamma = \mathbb{Z}[\pi_1 W]$, and let $C_* = C_*(W; \mathbb{Z}[\pi_1 W])$ be the chain complex of $\widetilde{W}$, considered as a complex of free left Γ-modules. Then $H_i(W;\Gamma) = H_i(C_*)$ is $H_i(\widetilde{W})$, with the natural Γ-module structure, for all i. The *equivariant cohomology* of $\widetilde{W}$ is defined in terms of the cochain complex $C^* = Hom_\Gamma(C_*, \Gamma)$, which is naturally a complex of right modules. Let $\overline{C^q}$ be the left Γ-module obtained via the canonical anti-involution of Γ, defined by $g \mapsto g^{-1}$ for all $g \in \pi_1 W$, and let $H^j(W;\Gamma) = H^j(\overline{C^*})$. We use similar notation for pairs of spaces. If W is a 4-manifold with boundary then equivariant Poincaré-Lefshetz duality gives isomorphisms $H_i(W;\Gamma) \cong H^{4-i}(W, \partial W; \Gamma)$ and $H^j(W;\Gamma) \cong H_{4-j}(W, \partial W; \Gamma)$, for all $i, j \leqslant 4$.

The *Wang sequences* for homology and cohomology (with coefficients R) associated to an infinite cyclic covering are the long exact sequences corresponding to the coefficient module sequence

$$0 \to R\Lambda \to R\Lambda \to R \to 0.$$

In Chapter 8 we shall use such sequences in the context of group (co)homology, when a group G has a normal subgroup K such that $G/K \cong \mathbb{Z}$.

Let X and Y be connected cell complexes, and let R be a $\mathbb{Z}[\pi_1 X]$-module. A map $f : Y \to X$ is a *R-homology isomorphism* if $\pi_1 f$ is an epimorphism and f induces isomorphisms on homology with local coefficients (induced from) R. A cobordism W with $\partial W = M_1 \sqcup M_2$ is an *H-cobordism over R* (or an R-homology cobordism) if the inclusion of M_i into W is an R-homology isomorphism, for $i = 1, 2$.

Let $\mathbb{F}$ be a prime field ($\mathbb{Q}$ or $\mathbb{F}_p$, where p is a prime). If G is a group the Massey product structures for classes in $H^1(G;\mathbb{F})$ are closely related to the rational and p-lower central series of G. We shall only use the following special case. Let a, b, c be classes in $H^1(G;R)$ represented by 1-cycles α, β and γ. If $a \cup b = b \cup c = 0$ then the Massey triple product $\langle a, b, c\rangle$ in $H^2(G;R)$ is the class represented by the 2-cycle $e_{\alpha,\beta}\gamma + \alpha e_{\beta,\gamma}$, where $e_{\alpha,\beta}$ and $e_{\beta,\gamma}$ are 1-chains such that $\partial^1(e_{\alpha,\beta}) = \alpha\beta$ and $\partial^1(e_{\beta,\gamma}) = \beta\gamma$ as 2-cocycles. (See [**Ma68, Dw75**] and also [**AIL**, Chapter 12].)

We shall usually omit the coefficients $\mathbb{Z}$ for integral homology, cohomology or Betti numbers.

1.6. Manifolds, Knots and Links

Let T be the torus, $T_g = \#^g T$ the closed orientable surface of genus $g \geqslant 0$, Kb the Klein bottle and $\#^c\mathbb{RP}^2$ the closed non-orientable surface with $c \geqslant 1$ cross-caps. In Chapter 6 we shall also use the notation $P_\ell = S^1 \cup_\ell e^2$ for the pseudo-projective plane with fundamental group $\mathbb{Z}/\ell\mathbb{Z}$. If B is a 2-orbifold then $|B|$ is the underlying surface. A 2-*handlebody* is a (smooth) 4-manifold constructed by adding 1- and 2-handles to D^4. If M is an n-manifold $M_o = \overline{M \setminus D^n}$ is the bounded manifold obtained by deleting a small open n-disc.

An embedding j of an m-manifold M in an $(m+k)$-manifold N is *locally flat* if each point $x \in M$ has a neighbourhood $U \subset M$ such that $j(U)$ has a product neighbourhood $V \cong U \times (-1,1)$ in N. (Note that PL embeddings of 3-manifolds into 4-manifolds are always locally flat, in the stronger PL sense, since the PL Schoenflies Theorem holds for PL embeddings of S^{n-1} in S^n when $n \leqslant 3$. We shall not need this observation.)

An *involution* of a topological space is a self-homeomorphism of order 2.

An n-sphere S^n is a *twisted double* if $S^n \cong W \cup_\theta W$, where W is a compact n-manifold with connected boundary and θ is a self-homeomorphism of ∂W.

If L is a link in S^3 with open regular neighbourhood $n(L)$ then $X(L) = S^3 \setminus n(L)$ is the link exterior, $\pi L = \pi_1 X(L)$ is the link group, and $M(L)$ is the 3-manifold obtained by 0-framed surgery on L. The order of $H_1(M(K);\Lambda)$ is generated by the Alexander polynomial of K.

An m-component link L in S^3 is *slice* if it bounds a set of m disjoint 2-discs properly embedded in D^4. (It is smoothly slice if the discs are also smoothly embedded.) It is a *ribbon link* if there is a map $R : mD^2 \to S^3$ which is locally an embedding, and whose only singularities are transverse double points, the double point sets being a disjoint union of intervals, and such that $R|_{m\partial D^2}$ is an embedding with image L. The map R may be homotoped *rel* ∂ to a proper embedding in D^4, and so every ribbon link is slice. (See [**Rol**, page 225] or [**AIL**, Chapter 1].)

A knot K in S^3 is *homotopically ribbon* if it bounds a 2-disc $D \subset D^4$ with an open product neighbourhood $n(D)$ such that the inclusion of $M(K) = \partial D^4 \setminus n(D)$ into $D^4 \setminus n(D)$ induces an epimorphism on fundamental groups [**CG83**]. Ribbon knots are homotopically ribbon. It is *doubly slice* if it is the transverse intersection of the equatorial S^3 in S^4 with an unknotted embedding of S^2 in S^4.

A *homology handle* is a 3-manifold M such that $H_1(M) \cong \mathbb{Z}$.

LEMMA 1.1. *Let M be a homology handle. Then there is a $\mathbb{Z}$-homology isomorphism $h : M \to S^2 \times S^1$.*

PROOF. Since $K(\mathbb{Z},1) \simeq S^1$ and $K(\mathbb{Z},2) = \mathbb{CP}^\infty$ may be constructed by adding cells of dimension $\geqslant 4$ to S^2, there are maps $f : M \to S^1$ and $g : M \to S^2$ such that $f^*\iota_1$ generates $H^1(M)$ and $g^*\iota_2$ generates $H^2(M)$. Let $h = (g,f) : M \to S^2 \times S^1$. Then h induces an isomorphism of cohomology rings. In particular, it is a $\mathbb{Z}$-homology isomorphism. □

Our principal references for knots and links are [**Rol**] and [**AIL**], and for the Kirby calculus [**GS**]. The notation for knots and links is as in [**Rol**], augmented by the symbols U and mU for the unknot and m-component trivial link, respectively.

1.7. Duality Pairings for 3- and 4-Manifolds

Let M be a closed connected orientable 3-manifold with fundamental group π, and let $\beta = \beta_1(M;\mathbb{Q})$. Let τ_M be the torsion subgroup of $H_1(M)$. Then Poincaré duality determines a linking pairing $\ell_M : \tau_M \times \tau_M \to \mathbb{Q}/\mathbb{Z}$. This is symmetric, bilinear, and non-singular in the sense that the adjoint function $\widetilde{\ell_M} : m \mapsto \ell(-,m)$ defines an isomorphism from τ_M to $Hom(\tau_M,\mathbb{Q}/\mathbb{Z})$. (Open 3-manifolds also have well-defined torsion linking pairings, but these may be singular.)

The linking pairing ℓ_M may be described as follows. Let w, z be disjoint 1-cycles representing elements of τ_M and suppose that $mz = \partial C$ for some 2-chain C which is transverse to w and some non-zero $m \in \mathbb{Z}$. Then $\ell_M([w],[z]) = (w \bullet C)/m \in \mathbb{Q}/\mathbb{Z}$. There is a dual formulation, in terms of cohomology. Let $\beta_{\mathbb{Q}/\mathbb{Z}} : H^1(M;\mathbb{Q}/\mathbb{Z}) \to H^2(M)$ be the Bockstein homomorphism associated with the coefficient sequence

$$0 \to \mathbb{Z} \to \mathbb{Q} \to \mathbb{Q}/\mathbb{Z} \to 0,$$

and let $D : H_1(M) \to H^2(M)$ be the Poincaré duality isomorphism. Then ℓ_M may be given by the equation

$$\ell_M(w,z) = (D(w) \cup \beta_{\mathbb{Q}/\mathbb{Z}}^{-1} D(z))([M]) \in \mathbb{Q}/\mathbb{Z}.$$

There are analogous pairings on covering spaces of M. In particular, if $\phi : \pi_1 M \to \mathbb{Z}$ is an epimorphism with associated covering space M_ϕ, and t is a generator t for the covering group, then the homology modules $H_1(M_\phi;R)$ are finitely generated $R\Lambda$-modules. Let $H_i(M;\mathbb{Q}\Lambda) = H_i(M_\phi;\mathbb{Q})$ and $H_i(M;\mathbb{Q}(t)) = \mathbb{Q}(t) \otimes_\Lambda H_i(M;\mathbb{Q}\Lambda)$. Let B be the $\mathbb{Q}\Lambda$-torsion submodule of $H_1(M;\mathbb{Q}\Lambda)$. Then equivariant Poincaré duality and the universal coefficient theorem together define a pairing $b : B \times B \to \mathbb{Q}(t)/\mathbb{Q}[t,t^{-1}]$, which is called the *Blanchfield pairing* associated to the covering. This pairing is non-singular and hermitian with respect to the involution sending t to t^{-1}. (See [**AIL**, Chapter 2].) Such a pairing is *neutral* if the underlying $\mathbb{Q}\Lambda$-torsion module has a submodule which is its own annihilator with respect to the pairing.

When ϕ corresponds to a fibre bundle projection from M to S^1 with fibre F a closed surface b_ϕ is equivalent to the isometric structure given by the intersection pairing I_F on $H_1(F;\mathbb{Q})$, together with the isometric action of $\mathbb{Z}$. (See [**Lit84**, Appendix A].)

Two such pairings are *Witt-equivalent* if they become isomorphic after addition of suitable neutral pairings. The *Witt group* of isometric structures on finite dimensional $\mathbb{Q}$-vector spaces is the set $W_+(\mathbb{Q}(t)/\mathbb{Q}\Lambda)$ of Witt equivalence classes of such pairings, with the addition induced by direct sum of pairings. (See [**Neu**].)

The primary manifestation of duality in dimension 4 is the *intersection pairing*. Let W be a compact oriented 4-manifold, and let $H = H_2(W;\mathbb{Z})/(torsion)$. Then every class $\alpha \in H_2(W;\mathbb{Z})$ is represented by a closed oriented surface F_α in M. Given two such classes α and β, we may assume that F_α and F_β intersect transversely. For each $P \in F_\alpha \cap F_\beta$ let $\varepsilon_P = 1$ if the combined orientations of the surfaces agree with the orientation of M, and let $\varepsilon_P = 1$ otherwise. Then

$$\alpha \bullet_W \beta = \Sigma_{P \in F_\alpha \cap F_\beta} \varepsilon_P$$

defines a pairing $\bullet_W : H \times H \to \mathbb{Z}$. (We shall write just $\bullet$ when there is no ambiguity.) This pairing has a more homological description. The homomorphism $D : H^2(W;\mathbb{Z}) \to H_2(W,\partial W;\mathbb{Z})$ defined by cap product with a fundamental class $[W]$ is an isomorphism, by Poincaré duality. Let $f_* : H_2(W;\mathbb{Z}) \to H_2(W,\partial W;\mathbb{Z})$ be the natural homomorphism. Then

$$\alpha \bullet_W \beta = D^{-1}(f_*\alpha)(\beta).$$

This pairing is symmetric, and is non-singular if f_* is an isomorphism. In general, the radical of this intersection pairing is the image in H of $\mathrm{Ker}(f_*)$, the image of $H_2(\partial W)$. If W is a subset of S^4 then $\bullet_W = 0$.

A knot K in S^3 is *algebraically slice* if its Blanchfield pairing is neutral.

1.8. Surgery

Surgery is a tool for constructing $\mathbb{Z}[\pi]$-homology equivalences $f : P \to W$ and $\mathbb{Z}[\pi]$-homology cobordisms between such maps, where P and W are closed n-manifolds and $\pi = \pi_1 W$. If $n \geqslant 4$ we may improve "$\mathbb{Z}[\pi]$-homology equivalence" to "homotopy equivalence", and if also the s-cobordism theorem holds then surgery becomes a means for finding homeomorphisms. There are also versions for maps of pairs of manifolds. In general, there are obstructions to achieving such goals, and delicate embedding issues limit the present applicability of surgery in dimensions $\leqslant 4$ to a class of groups including all virtually solvable groups.

We shall give a brief, very utilitarian "black box" outline of the surgery exact sequence as it is used in Chapters 6–8. For more details see either the original source [**Wall**] or the more recent [**Ran**]. The extension of surgery methods to the 4-dimensional TOP case is covered in [**FQ**] and [**BKKPR**]. See also [**KT02**].

Let W be a compact, connected orientable n-manifold, where $n \geqslant 5$, and let $\pi = \pi_1 W$. The *structure set* $\mathcal{S}_{TOP}(W, \partial W)$ is the set of equivalence classes of simple homotopy equivalences $f : (P, \partial P) \to (W, \partial W)$, where P is a compact n-manifold and $f|_{\partial P}$ is a homeomorphism, and where two such maps f_1 and f_2 are equivalent if there is a homeomorphism $h : P_2 \to P_1$ such that $f_2 \simeq f_1 \circ h$. This is a pointed set with distinguished element id_W. The structure set is the central object of interest for surgery, and sits in a sequence

$$L_{n+1}(\mathbb{Z}[\pi]) \xrightarrow{\ \omega\ } \mathcal{S}_{TOP}(W, \partial W) \xrightarrow{\ \nu_W\ } \mathcal{N}(W, \partial W) \xrightarrow{\sigma_n(W,\partial W)} L_n(\mathbb{Z}[\pi])$$

where the *surgery obstruction groups* $L_{n+1}(\mathbb{Z}[\pi])$ and $L_n(\mathbb{Z}[\pi])$ are algebraically defined, $L_{n+1}(\mathbb{Z}[\pi])$ acts on $\mathcal{S}_{TOP}(W, \partial W)$ via ω, the set of *normal invariants* $\mathcal{N}(W, \partial W) = [W, \partial W; G/TOP, *]$ is a topologically defined abelian group and $\sigma_n = \sigma_n(W, \partial W)$ is a homomorphism. This sequence is exact in the sense that two elements of the structure set have the same image in $\mathcal{N}(W, \partial W)$ if and only if they are in the same orbit of ω, and the kernel of σ_n is the image of the normal invariant map ν_W. (We use the ring-theoretic notation of Ranicki rather than the original group-theoretic notation of Wall for the surgery obstruction groups, writing $L_n(\mathbb{Z}[\pi])$ rather than $L_n(\pi)$. In particular, the groups $L_n(\mathbb{Z})$ are the obstruction groups for 1-connected surgery. In the cases we consider all homotopy equivalences are simple, and so we have written L_n rather than L_n^s.)

In dimension 4 we must modify the definition. The s-cobordism structure set $\mathcal{S}^s_{TOP}(W, \partial W)$ is the set of equivalence classes of simple homotopy equivalences f as above, where simple homotopy equivalences f_1 and f_2 are equivalent if there is an s-cobordism Q *rel* ∂ between P_1 and P_2 and a map $F : (Q, \partial Q) \to (W, \partial W)$ which extends $f_1 \sqcup f_2$. There is again a sequence of pointed sets

$$\mathcal{S}^s_{TOP}(W, \partial W) \to \mathcal{N}(W, \partial W) \to L_4(\mathbb{Z}[\pi]).$$

We may identify $\mathcal{N}(W, \partial W)$ with $H^2(W, \partial W; \mathbb{F}_2) \oplus \mathbb{Z}$, and σ_4 is trivial on the image of ν_W and injective on the $\mathbb{Z}$ summand. Although we do not know whether a 4-dimensional normal map with trivial surgery obstruction must always be in the image of ν_W ("normally cobordant to a homotopy equivalence"), in our applications we shall always have a homotopy equivalence in hand. A more serious problem is that it is not clear how to define the action ω in general. *Ad hoc* arguments apply in some of the cases of interest to us. (See [**FMGK**, Chapter 6.2].)

If π is virtually solvable (or, more generally, is "good" in the sense of [**FQ**]) then the s-cobordism theorem holds, and so $\mathcal{S}^s_{TOP}(W, \partial W) = \mathcal{S}_{TOP}(W, \partial W)$, and we again have a 4-term surgery exact sequence as above. In this case the order of the structure set is bounded by the order of $H^2(W, \partial W; \mathbb{F}_2) \cong H_2(W; \mathbb{F}_2)$.

The 4-dimensional Poincaré Conjecture follows immediately from the exactness of the surgery sequence with $W \simeq S^4$ and $\pi = 1$, for then $L_5(\mathbb{Z}) = 0$, so ω is trivial, and $H^2(W; \mathbb{F}_2) = 0$, and so $\sigma_4(W)$ is injective. Hence $\mathcal{S}_{TOP}(S^4)$ has only one member; every homotopy 4-sphere is homeomorphic to S^4.

In dimension 3 we need a more drastic modification. The structure set is now defined in terms of $\mathbb{Z}[\pi]$-homology equivalences, and the equivalence relation is coarsened to $\mathbb{Z}[\pi]$-homology cobordism [**KT02**]. (We use this in Section 6.5.)

The account given thus far is appropriate for the comparison of manifolds of the same homotopy type. However, the theory also handles the question of the existence of manifolds within a given homotopy type. The relevant homotopy types are PD_n-complexes. These are finite cell complexes which satisfy a strong form of Poincaré duality (taking into account the action of the fundamental group on covering spaces), of formal dimension n. They have stable spherical fibrations, which in the manifold case are stabilizations of the sphere bundles associated to the stable normal bundle. There is a parallel notion of PD_n-pairs. We may extend the definitions of structure set and normal invariants by allowing $(W, \partial W)$ to be a PD_n-pair. If (X, M) is a PD_n-pair with boundary M a closed $(n-1)$-manifold and whose stable spherical fibration reduces to a TOP bundle, and if there is a normal invariant f with $\sigma_n(f) = 0$ then the surgery exact sequence above remains valid with (X, M) in place of $(W, \partial W)$, and $\mathcal{S}^s_{TOP}(X, M)$ is non-empty. In other words, there is an n-manifold W with $\partial W \cong M$ such that $(W, \partial W) \simeq (X, M)$. We shall only use this extension in Chapter 8, in the simplest case, with X contractible, to outline Freedman's Embedding Theorem for homology 3-spheres.

Two obvious issues are the determination of the surgery obstruction groups $L_n(\mathbb{Z}[\pi])$ and the surgery obstruction homomorphisms σ_n. The groups $L_n(\mathbb{Z}[\pi])$ are periodic in n, with period 4. When π is finite the calculation of these groups involves representation theory and algebraic number theory (as in [**Wa76**]), but when there is a finite $K(\pi, 1)$ complex it is expected that these groups are largely determined by the homology of π with coefficients in $L_0(\mathbb{Z}) = \mathbb{Z}$ and $L_2(\mathbb{Z}) = \mathbb{Z}/2\mathbb{Z}$. (This is roughly the content of the Novikov and Farrell-Jones Conjectures. See [**KL**].) The fundamental groups that we shall consider in the second half of this book are mostly of the latter type, and the main difficulty that we shall meet in attempting to apply surgery is in identifying the homotopy types of pairs (W, M).

1.9. The G-Signature Theorem

The Index Theorem is one of the major achievements of the past century. Like many great theorems, there are a number of proofs, emphasizing different aspects and using a variety of techniques. Fortunately we shall only need two very special cases, for a finite cyclic group G acting on a closed orientable surface and for involutions of a closed orientable 4-manifold, with "good" fixed point set. The G-Signature Theorem in these cases relates global invariants based on signatures of bilinear pairings derived from cup-product to properties of the action of G on the normal bundle of the fixed point set.

The intersection form (,) on the middle dimensional homology of a $2k$-manifold M induces a hermitean form ϕ on $H = H_k(M;\mathbb{C}) = \mathbb{C}\otimes H_k(M;\mathbb{Z})$ by the formulae $\phi(\alpha x, \beta y) = \alpha\overline{\beta}(x,y)$ for k even and $\phi(\alpha x, \beta y) = i\alpha\overline{\beta}(x,y)$ for k odd. If G is a finite group acting orientably on M then there is a G-equivariant orthogonal decomposition $H = H^+ \oplus H^- \oplus H^0$, where ϕ is positive definite on H^+, negative definite on H^- and zero on H^0. If $g \in G$ then we let

$$\mathrm{sign}(g, M) = tr(g_*|_{H^+}) - tr(g_*|_{H^-}).$$

The value $\sigma(M) = sign(1, M)$ is the *signature* of M. Clearly $|sign(g, M)| \leqslant \beta_k(M;\mathbb{C})$, for all $g \in G$.

Suppose first that $G = \mathbb{Z}/\sigma\mathbb{Z}$ acts on a closed orientable surface M, with finite fixed point set F. Then the G-Index Theorem gives

$$\mathrm{sign}(g, M) = -i\Sigma_{P\in F}\cot(\theta_P),$$

where g rotates a disc neighbourhood of the fixed point P through θ_P radians.

Suppose now that g is an involution of a closed connected 4-manifold M such that the fixed point set F is a finite disjoint of surfaces, and g acts linearly on neighbourhoods of points in F. The G-Signature Theorem then gives

$$\mathrm{sign}(g, M) = e(F),$$

where $e(F)$ is the normal Euler number, which is just the Euler number of $\partial N(F)$, considered as a circle bundle over F.

The G-Signature Theorem is due to Atiyah, Bott and Singer [**AB68, AS68**]. An account of the theorem for G finite and M of dimension 2 or 4 which is adapted to the needs and backgrounds of low-dimensional topologists may be found in [**Go86**]. The manifold and the action are assumed there to be smooth, but smoothness is required for the proofs only in the neighbourhood of the fixed point set.

1.10. Other Tools

We mention briefly several other notions which are perhaps not yet part of the standard background of geometric topologists, but which appear in minor roles in this book. The Andrews-Curtis moves are used in Theorem 2.8, but this result is not used elsewhere in the book. See [**HAMS**]. In Theorem 7.10 we refer to the Bass Conjectures [**Ba76**], in order to show that certain projective modules are 0. Profinite completion is used in Theorem 9.4. See [**Ser**].

We shall say a little more about L^2-(co)homology, which is invoked in Chapters 7 and 9. This was originally defined analytically, in terms of the L^2-completion of $C_*(\widetilde{X};\mathbb{R})$, with square-summable chains and cochains (and the lifts of the q-cells of X representing an orthonormal basis for the completion of $C_q(\widetilde{X};\mathbb{R})$). It has since been reformulated as (co)homology with twisted coefficients in the von Neumann algebra $\mathcal{N}(G)$. The L^2-Betti number $\beta_i^{(2)}$ is the von Neumann dimension of $H_i(X;\mathcal{N}(G))$, and is a non-negative real number if X is a finite complex. These Betti numbers are multiplicative under transition to finite covering spaces, and there is an L^2-Euler formula $\chi(X) = \Sigma(-1)^q\beta_q^{(2)}(X)$ for finite complexes. If X is an n-manifold (or, more generally, a PD_n-complex) then $\beta_{n-i}^{(2)}(X) = \beta_i^{(2)}(X)$ for all i. We shall use L^2-(co)homology primarily through the fact that if X is a finite 2-complex, $\chi(X) = 0$ and $G = \pi_1 X$ is infinite and $\beta_1^{(2)}(G) = 0$, then X is aspherical [**FMGK**, Theorem 2.4]. The class of groups G for which $\beta_1^{(2)}(G) = 0$ includes all amenable groups and all groups with a finitely generated infinite normal subgroup of infinite index. See [**Lück**].

CHAPTER 2

Invariants and Constructions

The key algebraic invariants for our purposes are the fundamental groups of the spaces involved, the torsion linking pairing on the 3-manifold and the Euler characteristic of the complementary regions. In this chapter we shall review the basic constraints on these invariants and describe the construction by 0-framed surgery on bipartitedly slice links, from which many of our examples derive. We also give an application of this construction by W. B. R. Lickorish, who showed that any two groups with balanced finite presentations and isomorphic abelianization could be realized as the fundamental groups of complementary regions of some embedding of a 3-manifold in S^4. In the final section we show that if a complementary region is not 1-connected then its fundamental group may be changed by 2-knot surgery.

2.1. Codimension-1 Embeddings

An embedding of one topological space A into another B is an injective continuous map $f : A \to B$ which induces a homeomorphism from A onto its image $f(A)$. This definition is too broad for the purposes of geometric topology, as it allows for pathologies such as Alexander's horned sphere [**Al24**]. We shall impose a condition that holds for all smooth embeddings.

DEFINITION. A *locally flat embedding* of an n-manifold M in an $(n+k)$-manifold N is an injective continuous map $j : M \to N$ such that for each $m \in M$ there is a neighbourhood U of $j(m)$ in N and a homeomorphism $h : U \to \mathbb{R}^{n+k}$ such that $h(U \cap j(M)) = \mathbb{R}^n \times \{O_k\}$, where O_k is the origin in $\mathbb{R}^k$. Two embeddings j and $\tilde{j}$ of M in N are *equivalent* if there are self-homeomorphisms ϕ of M and ψ of N such that $\psi j = \tilde{j}\phi$.

Our focus shall be primarily on embeddings of closed 3-manifolds in S^4, but we shall give a very brief overview of codimension-1 embeddings (i.e., $k = 1$) in spheres of other dimensions.

If a closed connected n-manifold M embeds in S^{n+1} then M must be orientable and the complement must have two components. These basic observations are consequences of Alexander duality and hold even if the embedding is not locally flat. In the smooth (or locally flat) case, there is a more geometric argument, since it is then clear that there are arcs transverse to M in S^{n+1}. If $S^{n+1} \setminus M$ were connected there would be a simple closed curve in S^{n+1} intersecting M transversally in one point, contradicting the fact that S^{n+1} is simply connected. It is easy to see that $S^{n+1} \setminus M$ has at most two components, and so it has exactly two components.

Thus M has a product neighbourhood $M \times (-1, 1)$, and so must be orientable. We shall call the closures of the components of $S^{n+1} \setminus M$ the *complementary regions* of the embedding, for brevity.

The only connected closed 1-manifold is the circle S^1, and embeddings of S^1 in S^2 are all equivalent, by the classical Schoenflies Theorem. (The proof actually shows that every embedding in the broader sense is equivalent to the equatorial embedding. See [**Si05**] for a modern account.) The study of embeddings of more than one copy of S^1 reduces to a simple combinatorial analysis of the lattice of ovals bounded by a finite family of disjoint simple closed curves in the plane.

Alexander showed that PL embeddings of S^2 in S^3 are standard, while locally flat embeddings of S^n into S^{n+1} are all equivalent (for each $n \geqslant 1$), by the generalized Schoenflies Theorem of Brown and Mazur. (Smooth embeddings of S^n into S^{n+1} are also smoothly standard if $n \geqslant 4$, but it remains an open problem whether this is so when $n = 3$.)

The Brown Collaring Theorem is a basic consequence of the existence of normal bundles in differential topology, and plays a role in the generalized Schoenflies Theorem. The present proof is due to R. Connelly [**Co71**]. An immediate consequence of this theorem is that every closed locally flat hypersurface $M \subset S^n$ has a product neighbourhood $M \times (-1, 1)$.

THEOREM 2.1 (Brown). *Let M be a compact manifold with non-empty boundary. Then ∂M has a collared neighbourhood $V \cong \partial M \times (0, 1]$.*

PROOF. Let $N = M \cup_{\partial M} \partial M \times \mathbb{R}_+$, where we identify $x \in \partial M$ with $(x, 0)$ in $\partial M \times \mathbb{R}_+$. We shall construct a homeomorphism of N which pushes the closed subspace M out onto $M \cup_{\partial M} \partial M \times [0, 1]$. Since ∂M is compact, it has a finite open cover $\mathcal{U} = \{U_1, \dots, U_n\}$ by subsets with product neighbourhoods. Thus there are embeddings $h_i : U_i \times [-1, 0] \to M$ such that $h_i(u_i, 0) = u_i$ for all $u_i \in U_i$ and $i \leqslant n$. These embeddings clearly extend to embeddings $H_i : U_i \times [-1, \infty) \to N$.

Since ∂M is a compact Hausdorff space there is a continuous partition of unity $\{\lambda_i\}_{i \leqslant n}$ subordinate to $\mathcal{U}$. For each $0 \leqslant a \leqslant 1$ let f_a be the self-homeomorphism of $[-1, \infty)$ given by $f_a(t) = t$ if $-1 \leqslant t \leqslant -\frac{1}{2}$, $f_1(t) = (1 - 2a)t + a$ if $-\frac{1}{2} \leqslant t \leqslant 0$ and $f_a(t) = t + a$ if $t \geqslant 0$. Let $P_i : N \to N$ be the self-homeomorphism which maps (u_i, t) to $(u_i, f_{\lambda_i(u_i)}(t))$, and which is the identity outside $U_i \times [-1, \infty)$, for all $u_i \in U_i$ and $i \leqslant n$. Then the composite $P = P_1 \circ P_2 \circ \dots \circ P_n$ is a homeomorphism such that $P(M) = M \cup_{\partial M} \partial M \times [0, 1]$. Clearly $V = P^{-1}(\partial M \times (0, 1])$ is as required. □

When $n = 2$, Alexander showed also that if M is the torus T then one of the complementary regions is $S^1 \times D^2$, and so the classification of embeddings of T reduces to knot theory. If $\chi(M) < 0$ then we can use pairwise connected sums to construct embeddings which are "knotted on both sides", but there are embeddings of a more complicated nature. See [**BPW19**].

Alexander's torus theorem has an analogue due to I. R. Aitchison: if $M = S^2 \times S^1$ then one of the complementary regions is $S^2 \times D^2$, and so the classification of embeddings of $S^2 \times S^1$ reduces to 2-knot theory. (See Theorem 7.2.)

We shall assume henceforth that $n = 3$. Unless otherwise stated, all 3-manifolds considered here shall be closed, connected and orientable, and we shall usually write "homology sphere" instead of "integral homology 3-sphere".

2.2. Embedding 3-Manifolds

A 3-manifold M embeds in $\mathbb{R}^4$ if and only if it embeds in S^4. Since our arguments are largely homological it is often more natural to assume that M embeds in a homology 4-sphere Σ. However, we shall focus on the standard case. An embedding $j : M \to S^4$ is *smoothable* if it is smooth with respect to some smooth structure on S^4, equivalently, if each complementary region is a 4-dimensional handlebody (i.e., may be constructed by attaching 1-, 2- and 3-handles to D^4). Although the embeddings that we shall construct are usually smooth embeddings in the standard 4-sphere, we wish to apply 4-dimensional topological surgery, and so henceforth *embedding* shall mean TOP locally flat embedding, unless otherwise qualified. A *Poincaré embedding* of M into a homology 4-sphere Σ is a homotopy equivalence $X \cup_M Y \simeq \Sigma$, where (X, M) and (Y, M) are PD_4-pairs.

Let $j : M \to S^4$ be an embedding, with complementary regions X and Y, and let j_X and j_Y be the inclusions of M into X and Y, respectively. Clearly X and Y are compact 4-manifolds, with boundary $\partial X = \partial Y = M$. We may assume X and Y are chosen so that $\chi(X) \leqslant \chi(Y)$. (On occasion, we shall use W for either of the complementary regions when the size of $\chi(W)$ is not relevant.) Let $\pi = \pi_1 M$, $\pi_X = \pi_1 X$ and $\pi_Y = \pi_1 Y$, and let $j_{X*} = \pi_1 j_X$ and $j_{Y*} = \pi_1 j_Y$. One of the subsidiary themes of this book is the interaction between $\chi(X)$ and $\pi_1 X$.

The symbols X, Y, π, π_X and π_Y shall have the above interpretations henceforth. We shall also abbreviate $\beta_1(M)$ as β, when the meaning is clear from the context.

We shall summarize the basic properties of the homology and cohomology of the complementary regions in the next lemma. One simple but important observation is that the natural homomorphism $H_2(X) \to H_2(X, M)$ is 0, since it factors through $H_2(S^4) \to H_2(S^4, Y)$ and an excision isomorphism, and similarly for $H_2(Y) \to H_2(Y, M)$. Equivalently, the intersection pairings are trivial on $H_2(X)$ and $H_2(Y)$. (See Theorem 8.9 for one use of this observation.)

LEMMA 2.2. *Suppose M embeds in S^4, with complementary regions X and Y. Then*

(1) $\chi(X) + \chi(Y) = 2$;
(2) $H_i(M; R) \cong H_i(X; R) \oplus H_i(Y; R)$, *for* $i = 1, 2$, *while* $H_i(X; R) = 0$ *for* $i > 2$, *for any simple coefficients R (and similarly for cohomology);*
(3) $H^1(X; R) \cong H_2(Y; R)$ *and* $H^2(X; R) \cong H_1(Y; R)$;
(4) $\beta = \beta_1(M) = \beta_1(X) + \beta_2(X)$;
(5) $1 - \beta \leqslant \chi(X) \leqslant \chi(Y) \leqslant 1 + \beta$ *and* $\chi(X) \equiv \chi(Y) \equiv 1 + \beta$ *mod* (2); *and*
(6) π_X *is homologically balanced.*

PROOF. The first assertion is clear, since $2 = \chi(S^4) = \chi(X) + \chi(Y) - \chi(M)$ and $\chi(M) = 0$. The Mayer-Vietoris sequence for $S^4 = X \cup_M Y$ with coefficients R

gives isomorphisms

$$H_i(M;R) \cong H_i(X;R) \oplus H_i(Y;R),$$

for $i = 1,2$, while $H_3(M;R) \cong R$ and $H_j(X;R) = H_j(Y;R) = 0$ for $j > 2$. Moreover, $H_2(X;R) \cong H^1(Y;R)$, by Poincaré-Lefshetz duality and excision (or by Alexander duality). Then $\beta = \beta_1(X) + \beta_2(X)$, so $\chi(X) = 1 + \beta - 2\beta_1(X)$, where $0 \leqslant \beta_1(X) \leqslant \beta$.

Since $H_2(\pi_X; F$ is a quotient of $H_2(X;F)$ for any field F, by Hopf's Theorem, the final assertion follows from (2) and (3). □

In particular, $H^1(M;R) \cong H^1(X;R) \oplus H^1(Y;R)$ has a basis consisting of epimorphisms which extend on one side or the other. If $\beta = 0$ then $\chi(X) = \chi(Y) = 1$, while if $\beta = 1$ then $\chi(X) = 0$ and $\chi(Y) = 2$.

The cohomology ring $H^*(M)$ is determined by the 3-fold product

$$\mu_M : \wedge^3 H^1(M) \to H^3(M)$$

and Poincaré duality. If we identify $H^3(M)$ with $\mathbb{Z}$ we may view μ_M as an element of $\wedge^3(H_1(M)/\tau_M)$. Every finitely generated free abelian group H and linear homomorphism $\mu : \wedge^3 H \to \mathbb{Z}$ is realized by some closed orientable 3-manifold [**Su75**]. (If $\beta \leqslant 2$ then $\wedge^3\mathbb{Z}^\beta = 0$, and so $\mu_M = 0$.)

LEMMA 2.3. *The cup product* 3*-form* μ_M *is* 0 *if and only if all cup products of classes in* $H^1(M)$ *are* 0*. Its restrictions to each of* $\wedge^3 H^1(X)$ *and* $\wedge^3 H^1(Y)$ *are* 0.

PROOF. Poincaré duality implies immediately that $\mu_M = 0$ if and only if all cup products from $\wedge^2 H^1(M)$ to $H^2(M)$ are 0. The second assertion is clear, since $H^3(X) = H^3(Y) = 0$. □

See [**Lev83**] for the parallel case of doubly sliced knots.

If $\mu_M \neq 0$ then $H^1(X)$ and $H^1(Y)$ are non-trivial proper summands of $H^1(M)$. However, if $\mu_M = 0$ this lemma places no condition on these summands.

The 3-form μ_M is 0 if and only if $\pi/\gamma_3^{\mathbb{Q}}\pi \cong F(\beta)/\gamma_3^{\mathbb{Q}}F(\beta)$ [**Su75**]. However, this is a rather weak condition. The next lemma gives a stronger result.

LEMMA 2.4. *If* $H_1(Y) = 0$ *then* $\pi/\gamma_k\pi \cong F(\beta)/\gamma_k F(\beta)$*, for all* $k \geqslant 1$.

PROOF. If $H_1(Y) = 0$ then $H_2(X) = 0$, and T must be 0, by the non-degeneracy of ℓ_M, so $H_1(M) \cong H_1(X) \cong \mathbb{Z}^\beta$. Let $f : \vee^\beta S^1 \to X$ be any map such that $H_1(f)$ is an isomorphism. Then j_X and f induce isomorphisms on all quotients of the lower central series [**St65**], and so $\pi/\gamma_k\pi \cong F(\beta)/\gamma_k F(\beta)$, for all $k \geqslant 1$. □

If M is the result of surgery on a β-component slice link L then it has an embedding with a 1-connected complementary region, and so this lemma applies. However there are such examples for which π does not have any epimorphisms onto $F(\beta)$. (See [**AIL**, Figure 8.1].)

There are parallel results for the rational lower central series and the p-central series, for primes p, with coefficients $\mathbb{Q}$ and $\mathbb{F}_p$, respectively. In particular, if $\beta_1(Y) = 0$ then $\pi/\gamma_k^{\mathbb{Q}}\pi \cong F(\beta)/\gamma_k^{\mathbb{Q}}F(\beta)$, for all $k \geqslant 1$.

The diagram of fundamental groups determined by the inclusions of M into X and Y and of X and Y into S^4 is a push-out diagram by Van Kampen's Theorem.

$$\begin{array}{ccc} \pi & \xrightarrow{j_{X*}} & \pi_X \\ {\scriptstyle j_{Y*}}\downarrow & & \downarrow \\ \pi_Y & \longrightarrow & 1. \end{array}$$

The use of this observation in the proof of Aitchison's Theorem (Theorem 7.2 below) is formalized in the next lemma.

LEMMA 2.5. *If j_{X*} is a split monomorphism then $\pi_Y = 1$.*

PROOF. Let $\sigma : \pi_X \to \pi$ be a homomorphism such that $\sigma \circ j_{X*} = id_\pi$, and let $f_X = j_{Y*} \circ \sigma$. Then $f_X \circ j_{X*} = id_{\pi_Y} \circ j_{Y*}$ and so id_{π_Y} factors through the push-out group 1. Hence $\pi_Y = 1$. □

2.3. Hyperbolicity of the Linking Pairing

In the early paper [**Ha38**] W. Hantzsche observed that if M embeds in S^4 then the torsion subgroup of $H_1(M)$ is a direct double. This follows easily from the Universal Coefficient Theorem and Alexander duality. Let τ_M, τ_X and τ_Y denote the torsion subgroups of $H_1(M)$, $H_1(X)$ and $H_1(Y)$, respectively. Then $\tau_M \cong \tau_X \oplus \tau_Y$, by the Mayer-Vietoris argument, and $\tau_X \cong Ext(\tau_Y, \mathbb{Z}) = Hom(\tau_Y, \mathbb{Q}/\mathbb{Z})$, since $H_1(M) \cong H^2(Y)$. Hence τ_X is non-canonically isomorphic to τ_Y, and so $\tau_M \cong \tau_X \oplus \tau_X$.

Looking more closely at the role of duality, A. Kawauchi and S. Kojima strengthened this result [**KK80**, Lemma 6.1].

LEMMA 2.6 (Kawauchi-Kojima). *If a closed connected 3-manifold M embeds as a locally flat submanifold of S^4 then ℓ_M is hyperbolic.*

PROOF. Let $D_{X,M} : H_2(X, M) \to H^2(X)$ and $D_M : H_1(M) \to H^2(M)$ be the duality isomorphisms determined by an orientation for (X, M), and let $\delta_X : H_2(X, M) \to H_1(M)$ be the connecting homomorphism in the homology exact sequence for the pair. Then $j_X^* D_{X,M} = D_M \delta_X$.

It follows from the Mayer-Vietoris sequence that $H_i(M)$ maps onto $H_i(X)$, for $i = 1$ or 2, and hence that there is a short exact sequence

$$0 \to H_2(X, M) \xrightarrow{\delta_X} H_1(M) \xrightarrow{j_{X*}} H_1(X) \to 0.$$

Let $\tau_{X,M}$ and $\tau_{Y,M}$ be the torsion subgroups of $H_2(X, M)$ and $H_2(Y, M)$, respectively. Then $D_{X,M}$ restricts to give an isomorphism $\tau_{X,M} \cong Ext(\tau_X, \mathbb{Z}) = Hom(\tau_X, \mathbb{Q}/\mathbb{Z})$. (Similarly, D_M restricts to give $\tau_M \cong Hom(\tau_M, \mathbb{Q}/\mathbb{Z})$.) Hence $|\tau_M| = |\tau_{X,M}||\tau_X|$, and so the sequence of torsion subgroups

$$0 \to \tau_{X,M} \to \tau_M \to \tau_X \to 0$$

is also exact. (In particular, $\tau_{X,M}$ and $\tau_{Y,M}$ are the kernels of the induced homomorphisms from τ_M to $H_1(X)$ and $H_1(Y)$, respectively.) Since

$$\ell_M(\delta_X a, \delta_X b) = D_M(\delta_X b)(\delta_X a) = j_X^* D_{X,M}(b)(\delta_X a) = D_{X,M}(b)(j_{X*}\delta_X a) = 0,$$

for all $a, b \in \tau_{X,M}$, the image of $\tau_{X,M}$ under δ_X is self-annihilating. Similarly, $\delta_Y(\tau_{Y,M})$ is self-annihilating, and clearly τ_M is the direct sum of these two subgroups. □

D. B. A. Epstein showed that punctured lens spaces $L(2k, q)_o$ do not embed in S^4 [**Ep65**]. The following corollary extends this result. The proof here is taken from [**GL83**]. (Note that open 3-manifolds have well-defined linking pairings, which may be singular.)

COROLLARY 2.6.1. *Let N be a connected 3-manifold (possibly open) which is a locally flat submanifold of S^4. If $x \in H_1(N)$ has order 2^k for some $k \geqslant 1$ then $2^{k-1}\ell_N(x, x) = 0$.*

PROOF. We may assume that x and a null-homology of $2^k x$ are supported in a compact bounded codimension-0 submanifold $Q \subset N$. Let $M = \partial(Q \times I)$ be the boundary of a regular neighbourhood of Q in S^4. Then $M = DQ$ is the double of Q along its boundary, and $\tau_M \cong A \oplus B$, where A and B are self-annihilating with respect to ℓ_M. Let x' denote the image of x in τ_M, and suppose that $x' = x_A + x_B$, with $x_A \in A$ and $x_B \in B$. Then $2^k x_A = 2^k x_B = 0$, and $\ell_N(x, x) =$

$$\ell_Q(x, x) = \ell_M(x', x') = \ell_M(x_A, X_A) + 2\ell_M(x_A, x_B) + \ell_M(x_B, x_B) = 2\ell_M(x_A, x_B).$$

Hence $2^{k-1}\ell_N(x, x) = 2^k \ell_M(x_A, x_B) = 0$. □

Kawauchi had earlier given a related result for the Blanchfield pairing on M corresponding to an epimorphism from $\pi_1 M$ to $\mathbb{Z}$ which is the restriction of an epimorphism from $\pi_1 X$ to $\mathbb{Z}$ [**Ka77**, Theorem 4.2].

THEOREM (Kawauchi). *Let X be an orientable 4-manifold with connected boundary M, and let $\phi : \pi_1 X \to \mathbb{Z}$ be an epimorphism such that the restriction to the image of $\pi_1 M$ is also an epimorphism. Let X_ϕ and M_ϕ be the corresponding $\mathbb{Z}$-covering spaces, and let t be a generator of the covering group $Aut(X_\phi/X) \cong \mathbb{Z}$. If $t^2 - 1$ acts invertibly on $H_1(M_\phi; \mathbb{Q})$ and $H_2(X_\phi, M_\phi; \mathbb{Q})$ is a $\mathbb{Q}[t, t^{-1}]$-torsion module then the Blanchfield pairing on $H_1(M_\phi; \mathbb{Q})$ is neutral. Hence the order ideal of $H_1(M; \mathbb{Q}\Lambda) = H_1(M_\phi; \mathbb{Q})$ has a generator of the form $\Delta_\phi(t) = f(t)f(t^{-1})$, for some $f(t) \in \mathbb{Q}[t, t^{-1}]$.* □

We shall only use this result when $\beta = 1$. The epimorphism ϕ is then unique up to sign, since $H^1(X) \cong H^1(M) \cong \mathbb{Z}$. See Lemma 3.10 for a proof of this case.

2.4. Comparison of Embeddings

If two embeddings j and $\tilde{j}$ are isotopic through locally flat embeddings then they are ambient isotopic, by the Isotopy Extension Theorem [**FNOP**, Theorem 2.20]. Embeddings $j_0, j_1 : M \to S^4$ are *s-concordant* if they extend to an embedding of $M \times [0,1]$ in $S^4 \times [0,1]$ whose complementary regions are s-cobordisms *rel* ∂. We need this notion as it is not yet known whether 5-dimensional s-cobordisms are always products.

We shall say that an embedding has a group-theoretic property (e.g., abelian, nilpotent, restrained ...) if the groups π_X and π_Y have this property. If $j : M \to$

S^4 is an abelian embedding and all abelian embeddings of M in S^4 are equivalent to j, we shall say that j is *essentially unique*.

LEMMA 2.7. *If M and M' are $\mathbb{Z}$-homology cobordant 3-manifolds then M embeds in a homology 4-sphere if and only if M' embeds in a (possibly different) homology 4-sphere.*

PROOF. Suppose that M embeds in a homology 4-sphere Σ, with complementary regions X and Y, and W is a $\mathbb{Z}$-homology cobordism with $\partial W = M \sqcup M'$. Let $Z = W \cup_{M'} W$ be the union of two copies of W, identified along the boundary M'. A Mayer-Vietoris argument shows that $\Sigma' = X \cup_M W' \cup_M Y$ is a homology 4-sphere, and M' is clearly a locally flat submanifold of W'. □

We shall use a variation of this construction, taking into account the fundamental groups, in Theorem 6.12.

Let L be the $(2, 2k)$-torus link and $M = M(L)$. Then $\tau_M = H_1(M) \cong (\mathbb{Z}/k\mathbb{Z})^2$, and ℓ_M is hyperbolic, since M embeds in S^4. Taking connected sums shows that every hyperbolic linking pairing is realized by some 3-manifold which embeds in S^4. Every closed orientable 3-manifold is $\mathbb{Z}$-homology cobordant to a 3-manifold which is Haken and hyperbolic [**Liv81, My83**]. It then follows from Lemma 2.7 that every hyperbolic linking pairing is realized by a hyperbolic $\mathbb{Q}$-homology sphere which embeds in a ($\mathbb{Z}$-)homology 4-sphere. In Chapter 3 we shall show that every hyperbolic linking pairing on a finite abelian group of odd order is realized by a Seifert fibred 3-manifold which embeds in S^4.

2.5. Some Constructions of Embeddings

Any closed orientable 3-manifold M may be obtained by integrally framed surgery on some r-component link L in S^3, with $r \geqslant \beta$. We may assume that the framings are even [**Ka79**], and then after adjoining copies of the 0-framed Hopf link $Ho = 2_1^2$ (i.e., replacing M by $M \# S^3 \cong M$) we may modify L so that it is 0-framed. (If the component L_i has framing $2k \neq 0$ we adjoin $|k|$ disjoint copies of Ho and band-sum L_i to each of the $2k$ new components, with appropriately twisted bands.)

DEFINITION. A link L is *bipartedly trivial* (respectively, *ribbon* or *slice*) if it has a partition $L = L_+ \cup L_-$ into two sublinks which are each trivial links (respectively, ribbon or slice links).

The partition then determines an embedding $j_L : M \to S^4$, given by ambient surgery on an equatorial S^3 in $S^4 = D_+ \cup D_-$. We add 2-handles to these 4-balls along L_+ on one side and along L_- on the other, using sets of disjoint slice discs to achieve the ambient surgery. The complementary components have $\chi = 1 + 2s - r$ and $1 - 2s + r$. If L_+ and L_- are smoothly slice then j_L is smooth, and if they are trivial each complementary region has a natural Kirby calculus presentation, with 1-handles represented by dotting the components of one part of L and 2-handles represented by the remaining components of L, with framing 0 [**GS**]. Hence it is homotopy equivalent to a finite 2-complex, and its fundamental group has a presentation with generators corresponding to the meridians of the dotted circles and relators corresponding to the remaining components. We extend this Kirby calculus notation to allow the deleted discs corresponding to the dotted components to be a set of slice discs. (In our figures we shall distinguish the moieties by using thick and thin lines, with dots • on thin lines.)

The notation j_L is ambiguous, for if L has more than one component it may have several different partitions leading to distinct embeddings. Moreover, we must choose a set of slice discs for each of L_+ and L_-. If L_- or L_+ is non-trivial, but is obviously a ribbon link, we shall use the discs obtained by desingularizing the ribbon discs, and then finding a presentation for the fundamental group of the corresponding complementary region is also straightforward.

If L is itself a slice link then $\beta = r$ and there are embeddings of $M(L)$ realizing each value of $\chi(X)$ allowed by Lemma 2.2, including one with a 1-connected complementary region.

F. Quinn showed that any pair of connected finite 2-complexes $\{C, D\}$ which satisfy the conditions $H^2(C) \cong H_1(D)$ and $H^2(D) \cong H_1(C)$ deriving from Alexander duality (as in Lemma 2.2) may be realized up to homotopy as the complementary regions of a smoothable embedding of some 3-manifold [**Qu01**, Corollary 1.5]. (The case when D and D are acyclic was treated earlier [**Cr88, Hu90**]. See also [**Lic03**] and [**Liv03**].) Although we shall not use this result directly, it provides the right conceptual framework for some of our examples. W. B. R. Lickorish found a similar result with more algebraic hypotheses and a more explicit use of link presentations, for the cases with $\chi(C) = \chi(D)$, and we shall outline his argument [**Lic04**]. (The hypothesis in Theorem 2.8 that the groups have balanced presentations is equivalent to requiring that $\chi(C) = \chi(D)$ in Quinn's result.) We shall give the topological part of the argument, and refer to [**Lic04**] for the proof of the next lemma.

LEMMA. *Let $\mathcal{P}$ be a balanced finite presentation of a group G and B be a square presentation matrix for G^{ab}. Then $\mathcal{P}$ is Andrews-Curtis equivalent to a presentation $\mathcal{Q}$ for which the associated presentation matrix for G^{ab} is $\left(\begin{smallmatrix} B & 0 \\ 0 & I_r \end{smallmatrix}\right)$ for some $r \geqslant 0$.* □

The proof is based on the correspondence between Andrews-Curtis moves on a group presentation [**HAMS**, Chapter 1] and elementary matrix operations (plus block diagonal enlargements and their inverses) on the associated presentation for the abelianization.

THEOREM 2.8 (Lickorish-Quinn). *Let $\mathcal{P}_1$ and $\mathcal{P}_2$ be balanced presentations of groups G_1 and G_2 such that $G_1^{ab} \cong G_2^{ab}$. Then $S^4 = X_1 \cup X_2$ is the union of two codimension-0 submanifolds with $\pi_1 X_1 \cong G_1$, $\pi_1 X_2 \cong G_2$ and connected boundary $\partial X_1 = \partial X_2 = M$. Each of X_1 and X_2 has a handle structure consisting of one 0-handle, n 1-handles and n 2-handles, with the associated presentations for $\pi_1 X_1$ and $\pi_1 X_2$ being Andrews-Curtis equivalent to $\mathcal{P}_1$ and $\mathcal{P}_2$, respectively.*

PROOF. We shall begin by using Andrews-Curtis (AC) moves to modify the presentations. Abelianizing $\mathcal{P}_1 = \langle a_1, \ldots, a_n \mid r_1, \ldots, r_n \rangle$ determines an $n \times n$

integral matrix C which is a presentation matrix for G_1^{ab}. The transpose C^{tr} is also a presentation matrix for G_1^{ab}, by the Elementary Divisor Theorem. We shall view C^{tr} as a presentation matrix for G_2^{ab}. Then there is a presentation $\mathcal{P}_2^{AC} = \langle \alpha_1, \ldots, \alpha_{n+r} \mid \rho_1, \ldots, \rho_{n+r} \rangle$ for G_2 which is AC-equivalent to $\mathcal{P}_2$ and for which the corresponding presentation matrix for G_2^{ab} is the block diagonal matrix $D_2 = \left(\begin{smallmatrix} C^{tr} & 0 \\ 0 & I_r \end{smallmatrix}\right)$ **[Lic04**, Corollary 2.2]. Add r new generators $a_{n+1}, \ldots, a_{n+r}$ and r new relators $r_{n+1} = a_{n+1}, \ldots, r_{n+r} = a_{n+r}$ to $\mathcal{P}_1$, to get an AC-equivalent presentation for G_1 for which the corresponding presentation matrix for G_1^{ab} is $A = D_2^{tr}$.

Suppose that in each relator r_i there are $n_+^{i,j}$ occurrences of the generator a_j and $n_-^{i,j}$ occurrences of its inverse a_j^{-1}, and that in each ρ_j there are $\nu_+^{i,j}$ occurrences of α_i and $\nu_-^{i,j}$ occurrences of the symbol α_i^{-1}. Then $n_+^{i,j} - n_-^{i,j} = A_{i,j} = \nu_+^{i,j} - \nu_-^{i,j}$. If $e(i,j) = n_+^{i,j} - \nu_+^{i,j} > 0$, alter $\mathcal{P}_2^{AC}$ by changing ρ_j to $\rho_j(\alpha_i\alpha_i^{-1})^{e(i,j)}$. If $e(i,j) < 0$, alter $\mathcal{P}_1$ by changing r_i to $r_i(a_j a_j^{-1})^{-e(i,j)}$. In this way we may assume that $n_+^{i,j} = \nu_+^{i,j}$ and hence also $n_-^{i,j} = \nu_-^{i,j}$, for all i, j.

We now construct a link in S^3 as follows. Let $D_1, \ldots, D_{n+r}$ and $\Delta_1, \ldots, \Delta_{n+r}$ be mutually disjoint oriented discs. For each pair (i,j) with $1 \leqslant i, j \leqslant n+r$ let $H_+^{i,j}$ be a set of $n_+^{i,j}$ copies of the positive Hopf link (of two ordered, oriented components with linking number $+1$), and let $H_-^{i,j}$ be a set of $n_-^{i,j}$ copies of the negative Hopf link. Each of these Hopf links is to be in a (small) ball in which each of the two components bounds an oriented disc meeting the other component in one point. These balls are to be all mutually disjoint and disjoint from the original discs. Now join the boundary of Δ_i once to the first component of each link in $\bigcup(H_+^{i,j} \cup H_-^{i,j})$ with (long thin) bands. Do this in order around $\partial\Delta_i$ specified by the relator r_i. When a_j occurs in the relator, connect to the first component of one of the links in $H_+^{i,j}$, and when a_j^{-1} occurs in the relator, connect to the first component of one of the links in $H_-^{i,j}$. Similarly when $\alpha_i^{\pm 1}$ occurs in ρ_j, connect to the first component of one of the links in $H_\pm^{i,j}$. For an occurrence of α_i any unused second component of any Hopf link in $H_+^{i,j}$ may be selected, and similarly for α_i^{-1}; it can easily be ensured that all the bands used are mutually disjoint and that they respect all orientations. (However, there are many ways of choosing the bands.) The numbers of links in $H_\pm^{i,j}$ are to be chosen so that each link in each $H_\pm^{i,j}$ has its first component banded to Δ_i and its second component banded to D_j. This banding process changes the original discs to two new collections $D'_1, \ldots, D'_{n+r}$ and $\Delta'_1, \ldots, \Delta'_{n+r}$, each of mutually disjoint discs, by adding to the original discs the bands and discs spanning the components of the Hopf links. Let $L_+ = \{\partial D'_i\}$ and $L_- = \{\partial \Delta'_j\}$ be the links given by the boundaries of these discs. Then $L = L_+ \cup L_-$ is bipartedly trivial.

Let $X_\pm$ be the handlebodies determined by giving $L_\pm$ the 0-framing and dotting $L_\mp$. The handle decomposition of $X_\pm$ gives rise to a presentation for $\pi_1 X_\pm$ in the standard way. We associate a generator a_i to the 1-handle corresponding to D'_i and a relator r_j to to 2-handle corresponding to Δ'_j. Then r_j has an entry $a_i^{\pm 1}$ for every signed point of $\partial\Delta'_j \cap D'_i$ taken in order along $\partial\Delta'_j$. The construction has been engineered so that X_- gives rise to $\mathcal{P}_1$ and X_+ gives rise to X_- and $\mathcal{P}_2^{AC}$. □

We note that the embeddings given by this theorem always have $\chi(X) = \chi(Y)$.

In particular, any two perfect groups with balanced presentations can be realized as π_X and π_Y for some embedding of a homology sphere in S^4. C. Livingston has given examples in which π_X is superperfect but has no balanced presentation [**Liv05**].

In the following lemma of A. Donald, the notion of doubly slice knot is extended to say that a link L in S^3 is *doubly slice* if it is the transverse intersection of the equator in S^4 with an unknotted embedding of S^2. (The embeddings studied in [**Do15**] are all smooth, but the argument also applies here.)

LEMMA 2.9. [**Do15**] *Let L be a doubly slice link in S^3, and let $B_n(L)$ be the n-fold cyclic branched cover of S^3, branched over L. Then $B_n(L)$ embeds in S^4.*

PROOF. Since L is doubly slice, the pair (S^3, L) sits inside (S^4, U), where U is an unknotted copy of S^2. The branched covering of S^3, branched over L, extends to a branched covering of S^4, branched over U, and $B_n(L)$ embeds in the covering space, as the preimage of S^3. Cyclic branched coverings of S^4 branched over U are again homeomorphic to S^4, and so $B_n(L)$ embeds in S^4. □

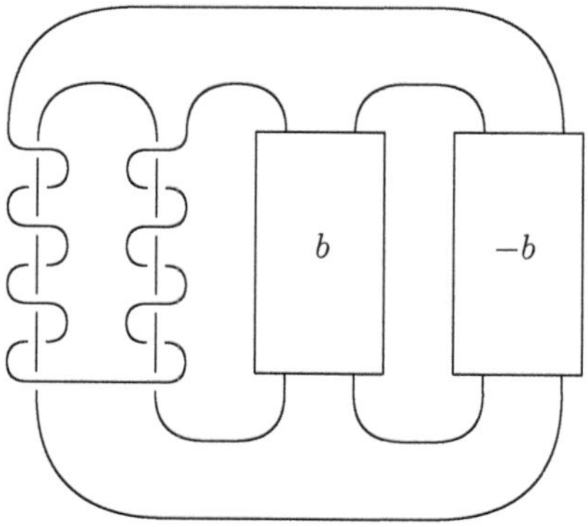

Figure 2.1 The pretzel link $P(-7, 7, b, -b)$.

The boxes contain 2-strand tangles with b twists in opposite directions (right over left, or left over right), as in the left half of the diagram.

Since cyclic branched coverings of S^4 branched over a smoothly unknotted 2-sphere are diffeomorphic to S^4, the same argument shows that if L is smoothly doubly slice then $B_n(L)$ embeds smoothly in S^4. Donald shows also that the pretzel links $P(a, -a, a)$, $P(a, -a, a, -a \pm 1)$ and $P(a, -a, a, -a)$ are all doubly slice, for any $a \in \mathbb{Z}$, while $P(a, -a, b, -b)$ is doubly slice if a and b are odd. In Figure 2.1 the letter b represents b half twists of a 2-strand tangle, with the minus sign representing reversal of the twist.

The "1-surgical embedding" of R. Budney and B. A. Burton gives smooth embeddings in smooth homotopy 4-spheres for 3-manifolds obtained by surgery on a slice link with all framings ± 1 [**BB22**]. (Note, however, that such 3-manifolds are homology spheres.)

2.6. Connected Sums

We may form ambient connected sums of embeddings as follows. Let $j : M \to \Sigma$ and $j' : M' \to \Sigma'$ be two embeddings of 3-manifolds in homology 4-sphere with complementary regions X, Y and X', Y', respectively. We shall assume that Σ and Σ' are oriented, and that M and M' have the induced orientation as the boundaries of X and X'. Choose small balls $D \subset \Sigma$ and $D' \subset \Sigma'$ such that $D \cap j(M)$ and $D' \cap j'(M')$ are each the equatorial 3-discs, and let $\Sigma_o = \overline{\Sigma \setminus D}$, $\Sigma'_o = \overline{\Sigma' \setminus D'}$, $M_o = \overline{M \setminus D}$ and $M'_o = \overline{M' \setminus D'}$. Let $h^+ : \partial D \to \partial D'$ and $h^- : \partial D \to \partial D'$ be orientation-reversing homeomorphisms such that $h^+(\partial D \cap X) = \partial D' \cap X'$ and $h^-(\partial D \cap X) = \partial D' \cap Y'$. Then $j|_{M_o} \cup j'|M'_o$ defines an embedding $j\#^+j'$ of $M\#M'$ in $\Sigma_o \cup_{h^+} \Sigma'_o$ and an embedding $j\#^-j'$ of $M\# - M'$ in $\Sigma_o \cup_{h^-} \Sigma'_o$. (If $\Sigma = \Sigma' = S^4$ then the new ambient homology 4-spheres are again copies of S^4.) The complementary regions are $X\natural X'$ and $Y\natural Y'$ for $j\#^+j'$, and $X\natural Y'$ and $Y\natural X'$ for $j\#^-j'$. Thus these two ambient connected sums are usually distinct.

If V and W' are complementary regions of j and j', respectively, then $\chi(V\natural W') = \chi(V) + \chi(W') - 1$ and $\pi_{V\natural W'} \cong \pi_V * \pi_{W'}$.

If M embeds then so does the punctured manifold M_o. Conversely, if M_o embeds then $M\#-M$ embeds as the boundary of a regular neighbourhood of M_o in S^4. If M_o and N_o each embed then so does $(M\#N)_o = M_o\natural N_o$; the converse is clear. Thus, for the question of whether M_o embeds we may assume M is irreducible. A 3-manifold which is a proper connected sum may embed in S^4 even though its indecomposable summands do not. The lens space $L = L(p,q)$ does not embed in S^4, since $H_1(L;\mathbb{Z})$ is not a direct double. However, if p is odd then L_o embeds smoothly as the fibre of a fibration of the complement of a twist spun 2-bridge knot [**Ze65**], and so $L\# - L$ embeds smoothly. (The cases with p even are excluded by [**Ep65**]. See also Corollary 2.6.1.)

2.7. Bi-epic Embeddings

DEFINITION. The embedding j is *bi-epic* if each of the homomorphisms $j_{X*} = \pi_1 j_X$ and $j_{Y*} = \pi_1 j_Y$ is an epimorphism.

If j is an embedding such that π_X and π_Y are nilpotent, then it is bi-epic, since $H_1(j_X)$ and $H_1(j_Y)$ are always epimorphisms.

LEMMA 2.10. *Let L be a bipartedly ribbon link. Then the embedding j_L constructed using the ribbon discs for each of the sublinks is bi-epic.*

PROOF. In this case π, π_X and π_Y are generated by images of the meridians of L. □

LEMMA 2.11. *The homomorphisms j_{X*} and j_{Y*} are both epimorphisms if and only if $j_\Delta = (j_{X*}, j_{Y*})$ is an epimorphism.*

PROOF. Let $K_X = \mathrm{Ker}(j_{X*})$ and $K_Y = \mathrm{Ker}(j_{Y*})$. If j_{X*} and j_{Y*} are epimorphisms then they induce isomorphisms $\pi/K_X \to \pi_X$ and $\pi/K_Y \to \pi_Y$. Hence $\pi/K_XK_Y \cong \pi_X/j_{X*}(K_Y)$ and $\pi/K_XK_Y \cong \pi_Y/j_{Y*}(K_X)$. Since $\pi_1(X \cup_M Y) = 1$, these quotients must all be trivial. If $g \in K_X$ and $h \in K_Y$ then $j_\Delta(gh) = (j_{X*}(h), j_{Y*}(g))$. Hence j_Δ is an epimorphism.

Conversely, if j_Δ is an epimorphism then so are its components j_{X*} and j_{Y*}. □

EXAMPLE 2.12. *Distinct embeddings derived from the same link.*

The link L obtained from the Borromean rings $Bo = 6^3_2$ by replacing one component by its $(2,1)$-cable and another by its $(3,1)$-cable may be partitioned as the union of two trivial links in three ways. The resulting three embeddings of $M(L)$ in S^4 each have $Y \simeq S^1 \vee 2S^2$, but the groups π_X have presentations $\langle a, b \mid [a, b^2]^3 \rangle$, $\langle a, c \mid [a, c^3]^2 \rangle$, and $\langle b, c \mid [b^2, c^3] \rangle$, respectively, and so are distinct. In the first two cases π has torsion, while in the third case X is aspherical. (None of these groups is abelian.) This example can obviously be generalized in various ways.

The homology sphere in Figure 2.2 is another example; the embedding determined by the link is bi-epic, but the 3-manifold also has an embedding with both complementary regions contractible. However, it is not clear whether there is such an embedding which derives from a 0-framed link representing the homology sphere.

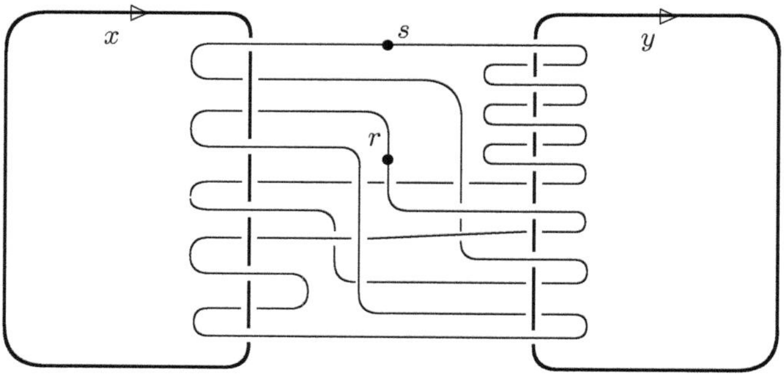

Figure 2.2 Both complements have group I^*.

If the embedding is nilpotent then j_Δ induces epimorphisms on all corresponding quotients of the lower central series, since $H_1(j_\Delta)$ is an isomorphism. However, these quotients are rarely isomorphic.

THEOREM 2.13. *If* $\pi/\gamma_3^{\mathbb{Q}}\pi \cong (\pi_X)/\gamma_3^{\mathbb{Q}}\pi_X) \times (\pi_Y)/\gamma_3^{\mathbb{Q}}\pi_Y)$ *then* $\chi(X) = 1 - \beta$ *or* $3 - \beta$.

PROOF. We use the fact that if G is a group then the kernel of cup product $\cup_G$ from $\wedge^2 H^1(G;\mathbb{Q})$ to $H^2(G;\mathbb{Q})$ is isomorphic to $\gamma_q^{\mathbb{Q}}G/\gamma_3^{\mathbb{Q}}G$. (See [**AIL**, §12.2].) Hence the rank of $\gamma_2^{\mathbb{Q}}G/\gamma_3^{\mathbb{Q}}G$ lies between $\binom{\beta_1(G)}{2} - \beta_2(G)$ and $\binom{\beta_1(G)}{2}$.

If the 2-step quotients $(G/\gamma_3^{\mathbb{Q}}G)$ are isomorphic then so are their commutator subgroups $\gamma_2^{\mathbb{Q}}\pi/\gamma_3^{\mathbb{Q}}\pi \cong (\gamma_2^{\mathbb{Q}}\pi_X/\gamma_3^{\mathbb{Q}}\pi_X) \times (\gamma_2^{\mathbb{Q}}\pi_Y/\gamma_3^{\mathbb{Q}}\pi_Y)$. Let $\gamma = \beta_1(X)$. Then $\gamma \geqslant \frac{\beta}{2}$, and $\beta_2(\pi) \leqslant \beta$, so the above bounds give

$$\binom{\beta}{2} - \beta \leqslant \binom{\gamma}{2} + \binom{\beta-\gamma}{2}.$$

This reduces to $\beta \geqslant \gamma(\beta - \gamma)$, and so either $\gamma \geqslant \beta - 1$ or $\beta = 4$ and $\gamma = 2$. In the latter case consideration of μ_M shows that the rank of $\gamma_3^{\mathbb{Q}}\pi/\gamma_3^{\mathbb{Q}}\pi$ is at least $3 \neq \binom{2}{2} + \binom{2}{2}$, so this cannot occur. Thus $\chi(X) = 1 + \beta - 2\gamma \leqslant 3 - \beta$. □

If j is any embedding with $\chi(X) = 1 - \beta$ then $H_2(j_\Delta; \mathbb{Q})$ is an epimorphism, and so j_Δ induces isomorphisms on all quotients of the rational lower central series. (If $H_1(Y) = 0$ then $\pi/\gamma_n\pi \cong \pi_X/\gamma_n\pi_X$, for all n.)

If F is a closed orientable surface then the embedding j of $M \cong F \times S^1$ as the boundary of a regular neighbourhood of the standard unknotted embedding of F in S^4 has $\chi(X) = 3 - \beta$ and j_Δ an isomorphism.

The cases when j_Δ is an isomorphism are quite rare.

LEMMA 2.14. *If j_Δ is an isomorphism then either $M \cong F \times S^1$ for some aspherical closed orientable surface F or $M \cong \#^r(S^2 \times S^1)$ for some $r \geqslant 0$.*

PROOF. If $\pi \cong \pi_X \times \pi_Y$ with π_X infinite and $\pi_Y \neq 1$ then $M \cong F \times S^1$ for some aspherical closed orientable surface F [**Ep61**]. If $\pi_Y = 1$ then j_{X*} is an isomorphism, and so π is a free group [**Da94**]. Hence $M \cong \#^r(S^2 \times S^1)$ for some $r \geqslant 1$. If π_X and π_Y are both finite and have non-trivial abelianization then their orders have a common prime factor p, and so π has $(\mathbb{Z}/p\mathbb{Z})^2$ as a subgroup, which is not possible. If they are finite perfect groups then so is π, and so $\pi \cong I^*$ or $\pi = 1$. Suppose that $\pi_X \neq 1$. Then $\pi \cong I^*$ and $\pi_Y = 1$, since I^* is not a non-trivial direct product. But I^* is not a free group, and so we may exclude this possibility. Thus if π is finite it must be trivial, and $M \cong S^3 = \#^0(S^2 \times S^1)$. □

Each of these manifolds has a bi-epic smooth embedding with j_Δ an isomorphism. (See Section 7.9.)

If π_X is a non-trivial proper direct factor of π then $\pi \cong \pi_1 F \times \mathbb{Z}$ for some closed orientable surface F, and so $M \cong F \times S^1$. In this case, either $F = S^2$ and $\pi_1 X \cong \mathbb{Z}$ or F is aspherical and $\pi_1 X \cong \pi_1 F$.

If π_X is a free factor of π then $M \cong M_X \# M'$, where $\pi_1 M_X \cong \pi_X$. The degree-1 collapse of M onto M_X induces an isomorphism $H_3(M) \cong H_3(M_X)$. Since $M = \partial X$ the images of $[M]$ and hence of $[M_X]$ in $H_3(\pi_X)$ are 0. Hence π_X is a free group (as in [**Da94**]). In particular, $\pi \cong \pi_X * \pi_Y$ only if π is itself a free group, and then $M \cong \#^\beta(S^2 \times S^1)$.

2.8. Modifying the Group

We may modify embeddings by 2-*knot surgery* on a simple closed curve in a complementary region, as follows. Let N_γ be a regular neighbourhood in X of a simple closed curve representing $\gamma \in \pi_X$. Then $\overline{S^4 \setminus N_\gamma} \cong S^2 \times D^2$ contains Y and M.

If K is a 2-knot with exterior $E(K)$ then $\Sigma = \overline{S^4 \setminus N_\gamma} \cup E(K)$ is a homotopy 4-sphere, and so is homeomorphic to S^4. Let $j_{\gamma,K}$ be the induced embedding of M in Σ. The complementary regions of $j_{\gamma,K}$ are $X_{\gamma,K} = \overline{X \setminus N_\gamma} \cup E(K)$ and Y. This construction applies equally well to simple closed curves in Y. We shall say that a 2-knot surgery is *proper* if γ is essential in X (or Y) and K is non-trivial.

When $M = S^2 \times S^1$ is embedded as the boundary of a regular neighbourhood of the trivial 2-knot, with $X = D^3 \times S^1$ and $Y = S^2 \times D^2$, the core $S^2 \times \{0\} \subset Y_1$ is K, realized as a satellite of the trivial knot. This construction gives all possible embeddings of $S^2 \times S^1$ in S^4 (up to equivalence), by Theorem 7.2.

Let t be the image of a meridian for K in the knot group $\pi K = \pi_1 E(K)$. If γ has infinite order in π_X then $\pi_1 X_{\gamma,K}$ is a free product with amalgamation $\pi_X *_{\mathbb{Z}} \pi K$; if it has finite order c then $\pi_1 X_{\gamma,K} \cong \pi_X *_{\mathbb{Z}/c\mathbb{Z}} \pi K/\langle\langle t^c \rangle\rangle$. (Note that if $K = \tau_c k$ is a non-trivial twist spin then we may assume that t^c is central, and so $\pi K/\langle\langle t^c \rangle\rangle \cong \pi K' \rtimes \mathbb{Z}/c\mathbb{Z}$.)

If $\gamma = 1$ then any simple closed curve representing γ is isotopic to one contained in a small ball, since homotopy implies isotopy for curves in 4-manifolds. Hence in this case 2-knot surgery does not change the topology of X.

It is well known that a nilpotent group with cyclic abelianization is cyclic. It follows that the natural projection of $\pi_1 X_{\gamma,K}$ onto π_X induces isomorphisms of corresponding quotients by terms of the lower central series. Thus we cannot distinguish these groups by such quotients. Nevertheless, we have the following result.

THEOREM 2.15. *If $\pi_X \neq 1$ then there are infinitely many groups of the form $\pi_1 X_{\gamma,K}$.*

PROOF. Suppose first that π_X is torsion-free and that $\gamma \neq 1$. If $\pi K \cong \mathbb{Z}/n\mathbb{Z} \rtimes \mathbb{Z}$ then $\pi_1 X_{\gamma,K} \cong \pi_X *_{\mathbb{Z}} \pi K$ is an extension of a torsion-free group by the free product of countably many copies of $\mathbb{Z}/n\mathbb{Z}$. Since $\mathbb{Z}/n\mathbb{Z} \rtimes \mathbb{Z}$ is the group of the 2-twist spin of a 2-bridge knot, for every odd n, the result follows.

If π_X has an element γ of finite order $c > 1$ then we use instead Cappell-Shaneson 2-knots. Let a be an integer, and let $f_a(t) = t^3 - at^2 + (a-1)t - 1$. If $a > 5$ the roots α, β and γ of f_a are real, and we may assume that $\gamma < \beta < \alpha$. Elementary estimates give the bounds

$$\frac{1}{a} < \gamma < \frac{1}{2} < \beta < 1 - \frac{1}{a} < a - 2 < \alpha < a.$$

If $A \in SL(3,\mathbb{Z})$ is the companion matrix of f_a then $\mathbb{Z}^3 \rtimes_A \mathbb{Z}$ is the group of a "Cappell-Shaneson" 2-knot K. The quotient $\mathbb{Z}^3/(A^c - I)\mathbb{Z}^3$ is a finite group of order the resultant $Res(f_a(t), t^c - 1) = (\alpha^c - 1)(\beta^c - 1)(\gamma^c - 1)$, where α, β and γ are the roots of $f_a(t)$. This simplifies to

$$\alpha^p + \beta^p + \gamma^p - (\alpha\beta)^p - (\beta\gamma)^p - (\gamma\alpha)^p = \alpha^p(1 - \beta^p - \gamma^p) + \varepsilon,$$

where $0 < \varepsilon < 2$. It follows easily from our estimates that $|Res(f_a(t), t^c - 1)| > a^{c-1}$, if $a > 3c$. Hence $\pi K/\langle\langle t^c \rangle\rangle$ is a finite group of order $> ca^{c-1}$. We then use the fact that finitely presentable groups have an essentially unique representation as the fundamental group of a graph of groups, with all vertex groups finite or one ended [**DD**, Proposition IV.7.4].) Thus if K and L are two such 2-knots such that $\pi K/\langle\langle t^c\rangle\rangle$ and $\pi L/\langle\langle t^c\rangle\rangle$ are finite groups of different orders, both greater than that of any of the finite vertex groups in such a representation of π_X, then $\pi_1 X_{\gamma,K} \not\cong \pi_1 X_{\gamma,L}$. □

The following observation can be construed as a minimality condition, since it shows that certain bi-epic embeddings cannot be obtained from other embeddings by non-trivial 2-knot surgery.

LEMMA 2.16. *Let M be a closed 3-manifold with an embedding $j : M \to S^4$, and let $\gamma \in \pi_{X(j)}$ have infinite order. Then the embedding $J = j_{\gamma,K}$ obtained from j by a proper 2-knot surgery using the 2-knot K and γ is not bi-epic. Moreover, $\pi_{X(J)}$ is not restrained, unless $\pi_{X(j)} \cong \mathbb{Z}$ and is generated by γ, in which case $\beta_1(M;\mathbb{Q}) = 1$ or 2, and $\pi_{X(J)} \cong \pi K$.*

PROOF. Let C be the subgroup of $\pi_{X(j)}$ generated by γ. Then

$$\pi_{X(J)} \cong \pi_{X(j)} *_C \pi K,$$

and the image of $\pi_1 M$ lies in $\pi_{X(j)}$.

The inclusion of $C \cong \mathbb{Z}$ into πK induces an isomorphism $C \cong \pi K^{ab}$. Since the 2-knot surgery is proper, $C \neq \pi K$, and so J is not bi-epic. Moreover, $\pi K' \neq 1$ and $[\pi K : C] > 2$, since $\pi K'$ has no quotient of order 2. Hence $\pi_{X(J)}$ has a non-cyclic free subgroup, unless γ generates $\pi_{X(j)}$ and so $C = \pi_{X(j)}$. If so, then $\chi(X(j)) = 0$ or 1 and so $\beta_1(M;\mathbb{Q}) = 1$ or 2. □

COROLLARY 2.16.1. *If j is bi-epic and π_X and π_Y are torsion-free then j is not the result of proper 2-knot surgery on any other embedding.* □

If γ has finite order c and $\pi K/\langle\langle t^c\rangle\rangle$ is non-abelian then the argument of Lemma 2.16 goes through, but otherwise it breaks down. For example, suppose that some power t^r with $r > 1$ is central. (This is so if K is an r-twist spin.) If $c \equiv 1$ *mod* (r) then the inclusion of C into $H = \pi K/\langle\langle t^c\rangle\rangle$ is an isomorphism, and so $\pi_{X(j)} \cong \pi_{X(J)}$. In this situation is $J = j_{\gamma,K}$ equivalent to j?

Although it is unlikely that there is a simple characterization of embeddings which are (not) obtained by 2-knot surgery on simpler embeddings, we can at least say that since the groups π_W are finitely presentable they are accessible, and so there is a bound on the possible decompositions along finite subgroups [**DD**, Chapter 4.7]. The theory of JSJ decompositions of groups provides a parallel result for decompositions along copies of $\mathbb{Z}$ [**GL**].

If $H_1(M) \neq 0$ then X is not simply connected, and so we may use 2-knot surgery to construct infinitely many embeddings with one complementary region Y and distinguishable by the fundamental groups of the other region. However if M is a homology sphere then X and Y are homology balls, and it may not be easy to decide whether π_X and π_Y are non-trivial. When $M = S^3$ the complementary regions are homeomorphic to the 4-ball D^4, by the generalized Schoenflies Theorem of Brown and Mazur. If $\pi_1 M \neq 1$, is there an homology 4-ball X with $M \cong \partial X$,

$\pi_X \neq 1$ and the normal closure of the image of $\pi_1 M$ in π_X being the whole group? If so, there is an embedding with one complementary region X and the other 1-connected.

Perhaps the simplest non-trivial example of a smooth embedding of an homology 3-sphere with neither complementary region 1-connected is given by the link in Figure 2.2. If we swap the 0-framings and the dots, we obtain a Kirby calculus presentation for Y. Since the loops r, s, x and y determine words $x^{-2}yxy$, $y^{-4}xyx$, $srsr^{-2}$ and $s^{-4}rsr$, respectively, π_X and π_Y have equivalent presentations, and $\pi_X \cong \pi_Y \cong I^*$, the binary icosahedral group.

CHAPTER 3

3-Manifolds with S^1-Actions

The class of Seifert manifolds is in many respects well understood, and has a natural parametrization in terms of Seifert data, and so we might expect criteria for embedding in terms of such data. We first review the notion of Seifert manifold and Seifert invariants. In this chapter we shall consider orientable Seifert manifolds which are Seifert fibred over orientable base orbifolds. These may also be described as (orientable) 3-manifolds of S^1-action type. (We shall consider 3-manifolds which are Seifert fibred over non-orientable base orbifolds in Chapter 4.) In Section 3.2 we show that Seifert manifolds bound compact manifolds constructed by plumbing disc bundles. The handle structures of these plumbed manifolds give framed link presentations for the bounding Seifert manifolds.

Motivated by the results of Hantzsche and Kawauchi and Kojima described in Section 2.3, we determine the torsion subgroup τ_M (for M of S^1-action type) in Section 3.3, and show that if M is Seifert fibred over $T_g(\alpha_1, \ldots, \alpha_r)$ and τ_M is a nonzero direct double then $|\varepsilon(M)|$ is determined by the cone point orders $\{\alpha_1, \ldots, \alpha_r\}$. A related calculation for M Seifert fibred over a non-orientable base orbifold plays an essential role in Chapter 4.

Our goal is to show that if M is a Seifert manifold with generalized Euler number $\varepsilon(M) = 0$, then skew symmetry of the Seifert data is necessary and sufficient for M to embed in S^4. We are partially successful. In Section 3.4 we show that if the Seifert data of M is skew-symmetric and all the cone point orders are odd then M embeds smoothly in S^4. Our main result (in Section 3.5) is that if the base orbifold is $S^2(\alpha_1, \ldots, \alpha_r)$, where all the cone point orders α_i are odd, and $\varepsilon(M) = 0$, then M embeds in S^4 if and only if the Seifert data is skew-symmetric. In Section 3.6 we outline briefly some apparent difficulties in extending this result. In particular, further conditions are needed if there are cone points of even order.

3.1. Seifert Manifolds

In this book a *Seifert manifold* is a closed orientable 3-manifold M with an orbifold S^1-fibration $p : M \to B$ over a 2-orbifold B. Since M is orientable, the *base* B has finitely many cone point singularities, and no reflector curves. If the surface $|B|$ underlying the base orbifold B is orientable and of genus g then $B = T_g(\alpha_1, \ldots, \alpha_r)$, where α_i is the order of the ith cone point. (The *cone point order* is also known as the multiplicity of the associated exceptional fibre.) The *Seifert data* for M is then a finite string of ordered pairs $S = ((\alpha_1, \beta_1), \ldots, (\alpha_r, \beta_r))$, where $(\alpha_i, \beta_i) = 1$, for all $1 \leqslant i \leqslant r$, and we shall write $M = M(g; S)$. If the base is non-orientable, so that $|B| = \#^c\mathbb{RP}^2$ for some $c > 0$, we shall write $M = M(-c; S)$. (We may refer to S as a Seifert data *set*, but the multiplicities of the pairs (α, β) are significant.)

Our notation is based on that of [**JN**]. (In particular, we do not assume that $0 < \beta_i < \alpha_i$.) We shall allow cone points of order $\alpha = 1$. Such points are non-singular, but are useful in that they allow a uniform notation which includes the Seifert fibrations associated to S^1-bundles. The reference [**JN**] also considers a *generalized Seifert manifold*, in which fibres of type $(0, 1)$ are allowed as well, corresponding to fibres fixed pointwise under a circle action. We use this definition, as a notational convenience only, in proving Theorem 4.4.

If $p : E \to F$ is an S^1-bundle with base a closed surface F and orientable total space E then $\pi_1 F$ acts on the fibre via $w = w_1(F)$, and such bundles are classified by an Euler class $e(p)$ in $H^2(F; \mathbb{Z}^w) \cong \mathbb{Z}$. If we fix a generator $[F]$ for $H_2(F; \mathbb{Z}^w)$ we may define the Euler number of the bundle by $e = e(p) \cap [F]$. (We may change the sign of e by reversing the orientation of E.) Let $M(g; (1, -e))$ and $M(-c; (1, -e))$ be the total spaces of the S^1-bundles with base T_g and $\#^c RP^2$ (respectively), and Euler number e. The *generalized Euler number* of a Seifert fibration $p : M \to B$ is

$$\varepsilon_S = -\Sigma_{i=1}^r \frac{\beta_i}{\alpha_i}.$$

There is an orientation- and fibre-preserving homeomorphism between any two Seifert manifolds with the same base orbifolds if and only if their Seifert data are equivalent under a finite sequence of the following operations

(1) add or delete any pair $(1, 0)$;
(2) replace each pair (α_i, β_i) by $(\alpha_i, \beta_i + c_i\alpha_i)$, where $\Sigma_{i=1}^r c_i = 0$;
(3) permute the indices.

Every Seifert data set is equivalent to one of the form $S = S' \cup \{(1, -e)\}$, where $S' = ((\alpha_1, \beta_1), \ldots, (\alpha_s, \beta_s))$ is *strict* Seifert data, with $0 < \beta_i < \alpha_i$ for all $i \leqslant s$. After reversing the orientation of the general fibre, if necessary, we may assume that $\varepsilon_S \geqslant 0$. In virtually all the cases of interest to us the Seifert fibration is unique, and so we shall usually write $\varepsilon(M)$ for ε_S.

Lens spaces, the manifolds $M(-1; (\alpha, \beta)) \cong M(0; (2, 1), (2, -1), (-\beta, \alpha))$ and the flat 3-manifold $M(-2; (1, 0)) \cong M(0; (2, 1), (2, 1), (2, -1), (2, -1))$ admit more than one Seifert fibration [**JN, Orl**].

Seifert manifolds $M = M(k; S)$ with $\varepsilon(M) = 0$ are generically $\mathbb{H}^2 \times \mathbb{E}^1$-manifolds, while those with $\varepsilon(M) \neq 0$ are generically $\widetilde{\mathbb{SL}}$-manifolds. The exceptions have virtually solvable fundamental groups and small Seifert data; the base orbifold has at most four singularities. They are the spherical manifolds S^3/G, the two $\mathbb{S}^2 \times \mathbb{E}^1$-manifolds $S^2 \times S^1$ and $\mathbb{RP}^3 \# \mathbb{RP}^3$, six (orientable) flat 3-manifolds and the $\mathbb{N}il^3$-manifolds. We shall settle the question of which of these embed in S^4 in Chapter 5.

The *fibred sum* of two Seifert manifolds $M = M(k; S)$ and $M' = M(k'; S')$ is defined as follows. Let N and N' be regular neighbourhoods of regular fibres in M and M', and let $h : \partial N \to \partial N'$ be an orientation-reversing, fibre-preserving homeomorphism. Then $M \sharp_f M' = (M \setminus int(N)) \cup_h (M' \setminus int(N')$ is Seifert fibred over the connected sum of the base orbifolds. In fact $M\sharp_f M' = M(k_\#; S \sqcup S')$, where $k_\# = k + k'$ if k and k' have the same sign, and $k_\# = -2k - k'$ if $k > 0$ and $k' < 0$. If $M' = M(0; (\alpha_i, \beta_i), (\alpha_i, -\beta_i))$ for some $(\alpha_i, \beta_i) \in S$ then $M\sharp_f M'$ is obtained from M by *expansion*, in the terminology of [**IM20**].

As B is the connected sum of $|B|$ and $S^2(\alpha_1, \dots, \alpha_r)$, the Seifert manifold M is a fibre sum $M = M(0; S)\sharp_f M(k; \emptyset)$. (If B is orientable then $M(g; \emptyset) \cong T_g \times S^1$.) If the Seifert data is given in normalized form $S = S' \cup \{(1, e)\}$ then we also have $M = M(0; S')\#_f M(k; (1, e))$, where $M(k; (1, e))$ is the total space of an S^1-bundle over F, with Euler number e.

If h is the image of the regular fibre in $\pi = \pi_1 M$ then the subgroup generated by h is normal in π, and $\pi^{orb}(B) \cong \pi/\langle h\rangle$. An orientable 3-manifold admits a fixed-point-free S^1-action if and only if it is Seifert fibred over an orientable base orbifold, and then h is central in π [**Orl**]. All Seifert manifolds considered in this chapter are of this type.

3.2. Plumbing and Link Presentations

In this section we shall consider Seifert manifolds as boundaries of smooth 4-manifolds. R. von Randow showed that the boundaries of plumbings of D^2-bundles over surfaces are precisely the graph manifolds [**vRan**]. (This includes all Seifert manifolds.) He also observed a connection with continued fractions, which we shall use to pass from Seifert data to plumbing data. W. Neumann subsequently gave criteria for two such plumbings to have homeomorphic boundaries and gave normal forms for representative plumbing graphs [**Neu81**]. The plumbing construction leads naturally to integrally framed link presentations.

We shall describe only the weighted graphs for Seifert manifolds and shall simplify the notation of [**Neu81**] accordingly. A *weighted graph* consists of a finite connected graph Γ such that each vertex has an integer weight. The *plumbed 4-manifold* $W(\Gamma)$ is built up as follows. For each vertex i of Γ, of valence d_i and weight e_i let F_i be the complement of d_i disjoint open discs in S^2, with a fixed orientation, and let E_i be a S^1-bundle over F_i. There is an "Euler number" obstruction in $\mathbb{Z}$ to extending a trivialization of the bundle over ∂F_i to a trivialization of E_i. We fix a trivialization which is compatible with the orientation and for which the Euler number is e_i. Whenever vertices i and j are connected by a $\pm$-edge in Γ we paste a boundary torus $T = S^1 \times S^1$ of E_i to a boundary torus of E_j by the map $J = \left(\begin{smallmatrix} 0 & 1 \\ 1 & 0 \end{smallmatrix}\right)$. These pasting maps flip base and fibre coordinates. Since they also reverse orientation, $W(\Gamma)$ inherits a compatible orientation from all of its pieces E_i. (In order to handle all graph manifolds and to establish his calculus, Neumann allows graphs with loops and multiple edges connecting the same pair of vertices. The edges then carry additional $\pm$ labels.)

We shall use the Hirzebruch-Jung formulation of (negative) continued fractions. (This differs from the standard formulation in the use of subtraction rather than addition.) If $a_1, \dots, a_n$ are integers $\geqslant 2$ we shall define the HJ continued fraction

$[a_1, \ldots, a_n]_{HJ}$ inductively by $[a_1] = a_1$ and

$$[a_1, \ldots, a_n]_{HJ} = a_1 - \frac{1}{[a_2, \ldots, a_n]_{HJ}} = a_1 - \cfrac{1}{a_2 - \cfrac{1}{a_3 - \cdots}}$$

if $n > 1$. The Euclidean algorithm may be used to show that every rational number $\frac{p}{q} > 1$ has an unique expression as such a continued fraction.

Suppose that $M = M(0; S)$, where $S = S' \cup \{(1, -e)\}$ is a strict Seifert data set. Then $M \cong \partial W(\Gamma_S)$, where Γ_S is derived from the Seifert data. We may assume that $S' = ((\alpha_1, \beta_1), \ldots, (\alpha_s, \beta_s))$, where $0 < \beta_i < \alpha_i$ for all $i \leqslant s$. Let $[a_1^i, \ldots, a_{n_i}^i]_{HJ}$ be the HJ continued fraction for α_i/β_i, for $i \leqslant s$. Then the weighted graph Γ_S associated to S is a tree, with one vertex of valence s and weight e, which we shall call the *central* vertex, and the complementary subgraph being the disjoint union of s linear graphs (of varying lengths) corresponding to these HJ continued fractions. Thus Γ_S has the shape illustrated in Figure 3.1.

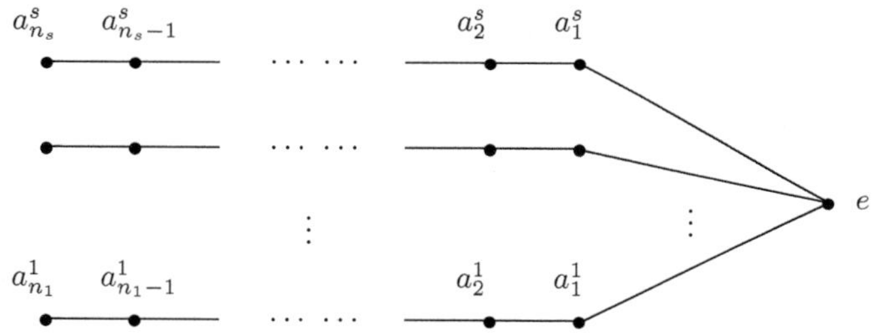

Figure 3.1 Γ_S.

The plumbed manifold $W(\Gamma_S)$ has a handle decomposition with one 0-handle and m 2-handles, where $m = 1 + \Sigma_{i=1}^s n_i$ is the number of vertices of Γ_S. The handle data may be presented via an integrally framed link, as in Figure 3.3 (ignoring the dotted components). In order to represent manifolds $W(k; S)$ with boundary $M(k; S)$, where $k \neq 0$, we must add g dotted components of type (a) in Figure 3.2 if $k = g > 0$ and c of type (b) if $k = -c < 0$.

Figure 3.2 Modifying the base orbifold.

Each component of type (a) corresponds to adding a copy of T to the base orbifold B, while those of type (b) correspond to adding copies of $\mathbb{RP}^2$ to B. In the latter case we must also change the framing of the central loop from $-e$ to $-e+2c$, to obtain a presentation for a 4-manifold $W(-c;S)$ with boundary $M(-c;S)$.

If we allow also weighted graphs which have several components we obtain similar link presentations for boundary-connected sums of plumbed 4-manifolds. The boundaries of such 4-manifolds are connected sums of Seifert manifolds. In particular, if we delete the central vertex and contiguous edges from Γ_S the resulting 4-manifold has boundary $\#^s L(\alpha_i, \beta_i)$, a connected sum of lens spaces.

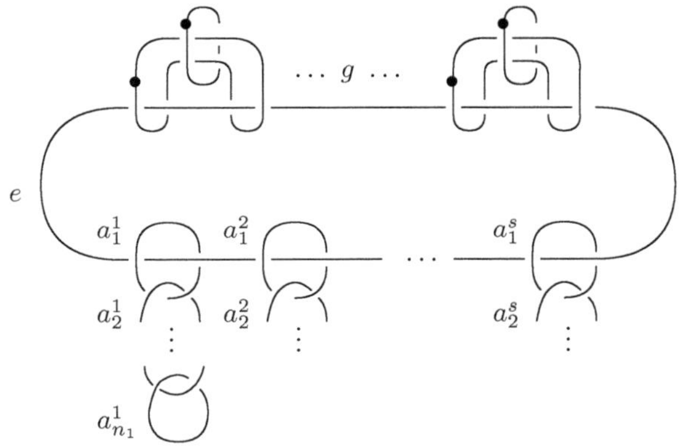

Figure 3.3 Link presentation for $M(g;S)$, with $g \geqslant 0$.

The rational surgery calculus may be used to modify Figure 3.2 to give a rational (Dehn surgery) presentation for the boundary [**Rol**, Sections 9G and 9H]. (Recall that the dotted components of the link presentation for the bounded 4-manifold become 0-framed components of the link presentation for the boundary [**GS**, Chapter 5.4].)

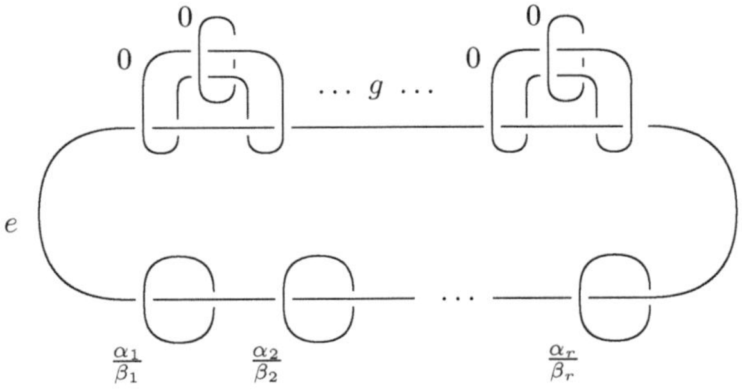

Figure 3.4 Link presentation for $M(0;S)$.

EXAMPLE 3.1. *The Poincaré homology sphere S^3/I^*.*

Let $S = ((2,1),(3,2),(5,4),(1,-2))$. Then $S^3/I^* = M(0;S)$, and Γ_S is the well-known $\mathbb{E}_8$ graph (see Figure 3.5), since $\frac{3}{2} = [2,2]_{HJ}$ and $\frac{5}{4} = [2,2,2,2]_{HJ}$.

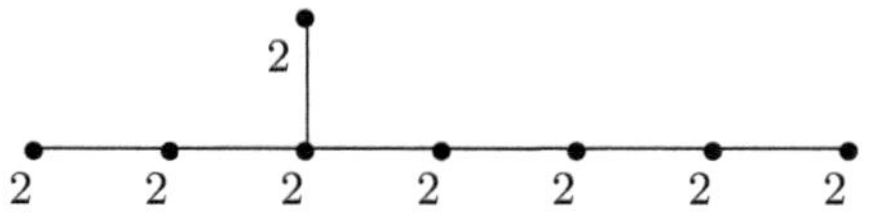

Figure 3.5 The $\mathbb{E}_8$ graph.

3.3. The Torsion Subgroup

In this section we shall describe the torsion subgroup τ_M in terms of the Seifert invariants of $M = M(g;S)$, when $g \geqslant 0$.

Let N_i be a regular neighbourhood of the ith exceptional fibre, and let B_o be a section of the restriction of the Seifert fibration to $\overline{M \setminus \cup\{N_i\}}$. Then B_o is homeomorphic to T_g with r open 2-discs deleted, and $\overline{M \setminus \cup\{N_i\}} \cong B_o \times S^1$. Let ξ_i and θ_i be simple closed curves on ∂N_i corresponding to the ith boundary component of B_o and a regular fibre respectively. The fibres are naturally oriented as orbits of the S^1-action. We may assume that B_o is oriented so that regular fibres have negative intersection with B_o, and the curves ξ_i are oriented compatibly with $\partial B_o = \Sigma\xi_i$. Then there are 2-discs D_i in N_i such that $\partial D_i = \alpha_i\xi_i + \beta_i\theta_i$. It follows from Van Kampen's Theorem that $\pi_1 M$ has a presentation

$$\langle a_1, b_1, \dots, a_g, b_g, q_1, \dots, q_r, h \mid \Pi[a_i, b_i]\Pi q_j = 1,\ q_i^{\alpha_i} h^{\beta_i} = 1,\ h \ central\rangle,$$

where $\{a_i, b_i\}$ are the images of curves in B_o which form a canonical basis for $|B|$, q_j is the image of ξ_j and h is the image of regular fibres such as θ_j.

THEOREM 3.2. *Let $M = M(g;S)$ be a Seifert manifold with orientable base orbifold. Then $H_1(M) \cong \mathbb{Z}^{2g} \oplus (\bigoplus_{i\geqslant 0}(\mathbb{Z}/\lambda_i\mathbb{Z}))$, where λ_i is determined by $\{\alpha_1, \dots, \alpha_r\}$ and is no-zero, for all $i > 0$. Moreover, λ_{i+1} divides λ_i, for all $i \geqslant 0$, and $|\varepsilon(M)|\Pi_{i=1}^n \alpha_i = \lambda_0 \Pi_{j\geqslant 1}\lambda_j$.*

PROOF. On abelianizing the above presentation for $\pi_1 M$, we see that $H_1(M) \cong \mathbb{Z}^{2g} \oplus \mathrm{Cok}(A)$, where A is the matrix

$$A = \begin{pmatrix} 0 & 1 & \dots & 1 \\ \beta_1 & \alpha_1 & \dots & 0 \\ \vdots & 0 & \ddots & \vdots \\ \beta_r & 0 & \dots & \alpha_r \end{pmatrix}.$$

Let $E_i(A)$ be the ideal generated by the $(r+1-i) \times (r+1-i)$ subdeterminants of A, and let δ_i be the positive generator of $E_i(A)$. Then $\Delta_0 = |det(A)| = |\varepsilon(M)|\Pi_{i=1}^n \alpha_i$. Since the elements of each row are relatively prime, Δ_i is the highest common factor of the $(r-i-1)$-fold products of distinct α_js, if $0 < i < r$, and $\Delta_i = 1$ if

$i \geqslant \max\{r-1,1\}$. (In particular, if $r > 2$ then $\Delta_{r-2} = \mathrm{hcf}(\alpha_1,\dots,\alpha_r)$.) Thus Δ_i depends only on $\{\alpha_1,\dots,\alpha_r\}$ and is nonzero, for all $i > 0$. If we set $\lambda_i = \Delta_i/\Delta_{i+1}$ for $i \geqslant 0$ then $\mathrm{Cok}(A) \cong \bigoplus_{i\geqslant 0}(\mathbb{Z}/\lambda_i\mathbb{Z})$, and λ_{i+1} divides λ_i, for all $i \geqslant 0$, by the Elementary Divisor Theorem. In particular, $|\varepsilon(M)|\Pi_{i=1}^n\alpha_i = \lambda_0\Pi_{j\geqslant 1}\lambda_j$. □

Note that $\tau_M \cong \tau_{M(0;S)}$ and the image of h is in τ_M if and only if $\varepsilon(M) \neq 0$.

COROLLARY 3.2.1. *If $\Delta_1 = 1$ then τ_M is cyclic, and $\tau_M = 0$ if and only if $\varepsilon(M) = 0$ or $\pm 1/\Pi_{i=1}^n\alpha_i$. If $\Delta_1 > 1$ then $\tau_M \neq 0$. Given $\{\alpha_1,\dots,\alpha_r\}$ such that $\Delta_1 > 1$, there is at most one value of $|\varepsilon|$ for which τ_M is a direct double.*

PROOF. As $\Delta_1 = \Pi_{i\geqslant 1}\lambda_i$ divides the order of τ_M, this group is non-zero unless $\Delta_1 = 1$. If $\varepsilon(M) \neq 0$ then $\tau_M \cong \bigoplus_{i\geqslant 0}(\mathbb{Z}/\lambda_i\mathbb{Z})$, and so is a direct double if and only if $\lambda_{2i} = \lambda_{2i+1}$ for all $i \geqslant 0$. In particular, $\varepsilon(M) = (\Delta_1)^2/\Pi_{i=1}^n\alpha_i\Delta_2$, and so is determined by $\{\alpha_1,\dots,\alpha_r\}$. If $\varepsilon(M) = 0$ then $\tau_M \cong \bigoplus_{i\geqslant 1}(\mathbb{Z}/\lambda_i\mathbb{Z})$, and so is a direct double if and only if $\lambda_{2i-1} = \lambda_{2i}$ for all $i > 0$.

Clearly these two systems of equations can both be satisfied only if $\lambda_i = 1$ for all $i > 0$ and $\lambda_0 = 0$ or 1, in which case $\tau_M = 0$. □

If M is Seifert fibred over a non-orientable base orbifold then $\tau_M \neq 0$, as we shall recall in Lemma 4.1.

It follows from Theorem 3.2 that a Seifert manifold M is a homology sphere if and only if $M = M(0;S)$ for some Seifert data S with $\varepsilon(M)\Pi_{i=1}^n\alpha_i = \pm 1$. In particular, $\mathrm{hcf}\{\alpha_i,\alpha_j\} = 1$ for all $i < j \leqslant r$. Conversely, if $r \geqslant 1$ and $\alpha_1,\dots,\alpha_r$ are pairwise relatively prime then an elementary arithmetic argument shows that the equation $\Sigma\frac{\beta_i}{\alpha_i} = \frac{1}{\Pi}$ has a solution for the numerators in integers, which is unique up to changing β_i by a multiple of α_i, for each $i \leqslant r$. Hence there is a unique homology sphere $\Sigma(\alpha_1,\dots,\alpha_r)$ which is Seifert fibred over $S^2(\alpha_1,\dots,\alpha_r)$. There is no reason to expect parity constraints for embedding such manifolds, since every homology sphere embeds in S^4 [**Fr82**].

If S' is strict Seifert data and $M = M(0;S',(1,-e))$ is a homology sphere then $e = \frac{1}{\Pi\alpha_i} + \Sigma_{i=1}^k\frac{\beta_i}{\alpha_i}$. It is easy to see that since e is an integer we must have $e < k$. This bound is best possible, by the following purely arithmetic lemma.

LEMMA 3.3. *Let $\alpha_1,\dots,\alpha_k$ be relatively prime integers greater than 1. If $k \geqslant 2$ then there are integers $\beta_1,\dots,\beta_k$ such that $0 < \beta_i < \alpha_i$ and $(\alpha_i,\beta_i) = 1$, for $i \leqslant k$, and*

$$\frac{1}{\Pi_{i=1}^k\alpha_i} + \Sigma_{i=1}^k\frac{\beta_i}{\alpha_i} = k-1.$$

PROOF. We shall induct on k. If $k = 2$ we may choose β_1 so that $0 < \beta_1 < \alpha_1$ and $1 + \beta_1\alpha_2 = m\alpha_1$, for some $m \in \mathbb{Z}$. Clearly $0 < m < \alpha_2$, and so $0 < \beta_2 = \alpha_2 - m < \alpha_2$. We then have

$$\frac{1}{\alpha_1\alpha_2} + \frac{\beta_1}{\alpha_1} + \frac{\beta_2}{\alpha_2} = 1.$$

Suppose the result holds for $k \leqslant j$. Let $\alpha_1,\dots,\alpha_{k+1}$ be relatively prime integers, and let $A = \alpha_k\alpha_{k+1}$. Let $\Pi = \Pi_{j=1}^{k+1}\alpha_j = (\Pi_{j=1}^{k-1}\alpha_j)A$. Then there are integers $\beta_1, \dots, \beta_{k-1}$, B such that $0 < \beta_i < \alpha_i$ and $(\alpha_i, \beta_i) = 1$, for all $i < k$, $0 < B < A$,$(A, B) = 1$ and $\frac{1}{\Pi} + \Sigma_{i=1}^{k-1}\frac{\beta_i}{\alpha_i} + \frac{B}{A} = k - 1$, by the inductive hypothesis.

If p, q are natural numbers then the additive subsemigroup of $\mathbb{N}$ generated by $\{p, q\}$ contains every integer $\geqslant (p-1)(q-1)$. Since α_k and α_{k+1} are relatively prime, and $A+B > \alpha_k\alpha_{k+1} > (\alpha_k - 1)(\alpha_{k+1} - 1)$, we may write $A + B = x\alpha_{k+1} + y\alpha_{k+2}$, where $x, y > 0$. Since $A + B < 2A$ and $(A, B) = 1$, we have $x < \alpha_k$ and $y < \alpha_{k+1}$, and $(x, \alpha_1) = (y, \alpha_2) = 1$. Let $\beta_k = x$ and $\beta_{k+1} = y$. Then $\frac{\beta_k}{\alpha_k} + \frac{\beta_{k+1}}{\alpha_{k+1}} = \frac{B}{A} + 1$, and so

$$\frac{1}{\Pi} + \Sigma_{i=1}^{k+1}\frac{\beta_i}{\alpha_i} = \frac{1}{\Pi} + \Sigma_{i=1}^{k-1}\frac{\beta_i}{\alpha_i} + \frac{B}{A} + 1 = k - 1.$$

This proves the inductive step, and hence the lemma. □

There is a similar criterion for a Seifert manifold to be a homology handle.

COROLLARY 3.3.1. *Let $M = M(g; S)$. Then there is a $\mathbb{Z}$-homology equivalence $f : M \to S^2 \times S^1$ if and only if $\varepsilon(M) = 0$ and* $\mathrm{hcf}\{\alpha_i, \alpha_j, \alpha_k\} = 1$ *for all $i < j < k \leqslant r$. If there are $r > 2$ cone points then no such map is a Λ-homology equivalence.*

PROOF. The first assertion follows from Theorem 3.5. If M is Seifert fibred then the base orbifold is orientable, since M is orientable and π/π' is torsion-free. Since $\varepsilon(M) = 0$, β is odd, and M is a mapping torus. If $r > 2$ then $\pi_1 M \not\cong \mathbb{Z}$, and so the infinite cyclic cover of M has positive genus. □

The elementary divisors λ_i may be determined more explicitly by localization. If p is a prime, an integer m has *p-adic valuation* $val_p(m) = v$ if $m = p^v q$, where p does not divide q.

COROLLARY 3.3.2. *Let p be a prime and let $v_i = val_p(\alpha_i)$, for $i \geqslant 1$. Assume that the indexing is such that $v_i \geqslant v_{i+1}$ for all i, and that τ_M is a direct double. Then*

(1) *if $\varepsilon(M) = 0$ then $v_{2j-1} = v_{2j}$ for all $j \geqslant 1$;*
(2) *if $\varepsilon(M) \neq 0$ then $val_p(\lambda_0) = v_3$, and $v_{2j} = v_{2j+1}$ for all $j \geqslant 2$.*

PROOF. It is clear from the theorem that $val_p(\Delta_i) = \Sigma_{j \geqslant i+2} v_j$, and so $val_p(\lambda_i) = v_{i+2}$, for all $i \geqslant 1$. If $\varepsilon(M) = 0$ then $v_1 = v_2$ follows immediately from the fact that $p^{v_2}\varepsilon(M)$ is an integer. If also $\bigoplus_{i \geqslant 1}(\mathbb{Z}/\lambda_i\mathbb{Z})$ is a direct double then $v_{2j-1} = v_{2j}$ for all $j \geqslant 2$.

Suppose now that $\varepsilon(M) \neq 0$. Then $val_p(\varepsilon(M))) \geqslant -v_1$, and $val_p(\varepsilon(M)) = -v_1$ if $v_1 > v_2$. Since $val_p(\Pi) = \Sigma_{i=1}^n v_i$ and $val_p(\Pi_{j \geqslant 1}\lambda_j) = val_p(\Delta_1)$, we have $val_p(\lambda_0) = v_1 + v_2 + val_p(\varepsilon(M)) \geqslant v_2$ (with equality if $v_1 > v_2$). If τ_M is a direct double we must have $val_p(\lambda_0) = v_2 = v_3$ and $v_{2j} = v_{2j+1}$ for all $j \geqslant 2$. □

We note here a related observation of Issa and McCoy.

LEMMA 3.4. [**IM20**, Lemma 4.4] *Let $M = M(g; \frac{\beta_1}{\alpha_1}, \dots, \frac{\beta_k}{\alpha_k}, (1, e))$, where $0 < \beta_i < \alpha_i$ for all i and $\varepsilon(M) > 0$, and suppose that $\tau_M \cong A \oplus A$ for some finite abelian group A. If the index set $\{1, \dots, k\}$ is partitioned into $n \leqslant e$ classes $\{C_i\}$ such that $\Sigma_{j \in C_i}\frac{\beta_j}{\alpha_j} \leqslant 1$ for all $i \leqslant n$ then $n = e$ and there is precisely one value of i for which $\Sigma_{j \in C_i}\frac{\beta_j}{\alpha_j} < 1$. For this value we have $1 - \Sigma_{j \in C_i}\frac{\beta_j}{\alpha_j} = \frac{1}{\mathrm{lcm}(a_1, \dots, a_k)}$.*

PROOF. The p-primary part of $\tau_M \cong A \oplus A$ is a direct double, for each prime p. We may assume the cone points are labelled so that the p-adic valuations of their orders are $v_k \geqslant v_{k-1} \geqslant \dots \leqslant v_1$. Let w be the p-adic valuation of $\varepsilon(M)$. The p-primary part of τ_M is then $\bigoplus_{i=3}^k (\mathbb{Z}/p^{v_i}\mathbb{Z}) \oplus \mathbb{Z}/p^v\mathbb{Z}$, where $v = v_1 + v_2 + w$, and $v \geqslant v_2$. Since there must be an even number of summands, either k is odd or $v_1 = 0$. Since the exponents v_i are increasing, this can be a direct double only if $v = v_1 = v_2$. But then $w = -v_1$ is the p-adic valuation of $\mathrm{lcm}(\alpha_1, \dots, \alpha_k)$.

This argument applies to each prime, and so

$$\varepsilon(M) = \frac{1}{\mathrm{lcm}(\alpha_1, \dots, \alpha_k)}.$$

If k is even then $\mathrm{hcf}(\alpha_1, \dots, \alpha_k) = 1$.

Now suppose that the index set $\{1, \dots, k\}$ is partitioned into $n \leqslant e$ classes $\{C_i\}$ such that $\Sigma_{j \in C_i} \frac{\beta_j}{\alpha_j} \leqslant 1$ for all $i \leqslant n$. Then we may write

$$\varepsilon(M) = e - n + \Sigma_{j=1}^n \sigma_j = (1 - \Sigma_{i \in C_j} \frac{\beta_i}{\alpha_i}) = \frac{1}{\mathrm{lcm}(\alpha_1, \dots, \alpha_k)},$$

where $\sigma_j = 1 - \Sigma_{i \in C_j} \frac{\beta_i}{\alpha_i}) \geqslant 0$ for all j. Then we must have $e - n \leqslant 0$ and so $n = e$. If $\sigma_j > 0$ then $\sigma_j \geqslant \frac{1}{\mathrm{lcm}(\alpha_1, \dots, \alpha_k)}$, and so $\sigma_h = 0$ for all $h \neq j$. □

The following bound is an immediate consequence. (Note that this result bounds e in terms of the number of singular fibres, rather than in terms of the full Seifert data for these singular fibres, as in Corollary 3.2.1.)

THEOREM 3.5. [**IM20**, Proposition 4.6] *Let $M = M(g; \frac{\beta_1}{\alpha_1}, \dots, \frac{\beta_k}{\alpha_k}, (1, -e))$, where $k > 1$, $0 < \beta_i < \alpha_i$ for all i and $\varepsilon(M) > 0$. If M embeds in S^4 then $e \leqslant k - 1$.*

PROOF. If M embeds in S^4 then τ_M is a direct double. If $e \geqslant k$ then the partition of $\{1, \dots, k\}$ into k singletons would violate the conclusion of Lemma 3.4, since $\frac{\beta_i}{\alpha_i} < 1$ for all $i \leqslant k$, and $k > 1$. □

For every $k \geqslant 3$ there are Seifert fibred $\mathbb{Z}$-homology 3-spheres of the form $M(0; \frac{\beta_1}{\alpha_1}, \dots, \frac{\beta_k}{\alpha_k}, (1, 1-k))$, by Lemma 3.3. Since these embed in S^4, by Freedman's result, the bound in this theorem is best possible.

Let ℓ be a linking pairing on a finite abelian group of order a power of an odd prime p. Then $\ell \cong \perp_{j=1}^t \ell_j$, where ℓ_j is a pairing on $(\mathbb{Z}/p^{k_j}\mathbb{Z})^{\rho_j}$ and $k_1 > k_2 > \dots > k_t \geqslant 1$. If ℓ is hyperbolic then each of the summands ℓ_j is also hyperbolic, and so the ranks ρ_j are all even. Let $r_1 = \frac{1}{2}\rho_1 + 1$ and $r_i = \frac{1}{2}\rho_i$ for $2 \leqslant j \leqslant t$, and let $S(p)$ be the concatenation of r_j copies of $((p^{k_j}, 1), (p^{k_j}, -1))$, for $1 \leqslant j \leqslant t$. Then $M = M(0; S(p))$ embeds in S^4, by Lemma 3.2, and $\ell_M \cong \ell$, since $\tau_M \cong \oplus_{j=1}^t (\mathbb{Z}/p^{k_j}\mathbb{Z})^{\rho_j}$ and ℓ_M is hyperbolic. This argument extends to show that every hyperbolic linking pairing on a finite abelian group of odd order is realized by some Seifert 3-manifold M with $\varepsilon(M) = 0$ and which embeds in S^4.

In Appendix A we analyze the linking pairings of orientable 3-manifolds with fixed-point-free S^1-actions. In particular, the linking pairing is even if and only if all even cone point orders α_i have the same 2-adic valuation, and if $|\tau_M|$ is even there are at least 3. (See Lemma A6 and Theorem A10.)

3.4. Skew Symmetry

The Seifert manifolds $M(g;S)$ with $g \geqslant 0$ and $\varepsilon(S) = 0$ which are known to embed in S^4 all have skew-symmetric Seifert data, in the following sense.

DEFINITION. The Seifert data S is *skew-symmetric* if it is equivalent to Seifert data of the form $((\alpha_1,\beta_1),\dots,(\alpha_{2s},\beta_{2s}))$, where $0 < \beta_{2i-1} < \alpha_{2i-1} = \alpha_{2i}$ and $\beta_{2i} = -\beta_{2i-1}$ for $1 \leqslant i \leqslant s$. Let $\frac{1}{2}S = ((\alpha_1,\beta_1),(\alpha_3,\beta_3)\dots,(\alpha_{2s-1},\beta_{2s-1}))$ be the subset of pairs with $\beta_i > 0$.

It is obvious that $\varepsilon(S) = 0$ if S is skew-symmetric. Our goal is to show that skew symmetry is necessary and sufficient for a Seifert manifold M with $\varepsilon(M) = 0$ to embed in S^4. In this section we shall demonstrate the sufficiency, for the cases with all cone point orders odd. Skew symmetry is a necessary condition if $g = 0$ and all cone point orders are odd, by Corollary 3.10.1, or if $g \geqslant 0$ and the embedding is smooth [**Do15**]. When there are cone points of even order the situation is not clear.

Every Seifert manifold $M(0;S)$ is the double branched cover of S^3, branched over a (Montesinos) link [**Mo73**]. If L is an m-component link and $M_2(L)$ is the associated double branched cover then $H^2(M_2(L);\mathbb{F}_2) \cong \mathbb{F}_2^{m-1}$. (This follows easily by the argument of [**AIL**, page 114], with coefficients $\mathbb{F}_2$ instead of $\mathbb{Z}$.) Hence if $M = M(0;S)$ and $H_1(M;\mathbb{Z})$ is finite of odd order then the branch set K is a knot. The 2-twist spin of K is a fibred 2-knot in S^4, with fibre the punctured double branched cover $M_2(k)_o$. The boundary of a regular neighbourhood of the fibre in S^4 is $M_2(K)\# - M_2(K)$. This is not Seifert fibred (except when $M_2(K) = S^3$) and so we shall modify the construction to get what we want.

If $S = ((\alpha_1,\beta_1),\dots,(\alpha_{2s},\beta_{2s}))$ is skew-symmetric Seifert data we shall let $\frac{1}{2}S = ((\alpha_1,\beta_1),(\alpha_3,\beta_3)\dots,(\alpha_{2s-1},\beta_{2s-1}))$ be the subset of pairs with $\beta_i > 0$. Let $D(\frac{1}{2}S)$ be the complement of an open tubular neighbourhood of a regular fibre of $M(0;\frac{1}{2}S)$. Then $\partial D(\frac{1}{2}S) \cong T$ and $M(0;S) = D(\frac{1}{2}S) \cup_\partial -D(\frac{1}{2}S)$ may be obtained by doubling $D(\frac{1}{2}S)$ along its boundary torus.

The following argument is based on [**IM20**, Proposition 7.8], with details from the earlier argument of [**CH98**, Lemma 3.1] which are needed for the extension in Lemma 3.8.

THEOREM 3.6. *Let S be skew-symmetric Seifert data for which all cone point orders α_i are odd. Then $M = M(0;S)$ embeds smoothly in S^4.*

We may assume that there is a 4-ball $B(M) = D^2 \times I \times [-\epsilon,\epsilon]$ embedded in S^4 such that $M \cap B(M) = \partial D^2 \times I \times [-\epsilon,\epsilon]$, and each circle $\partial D^2 \times (r,s)$ is a regular fibre of M.

PROOF. Let $N = M(0;\frac{1}{2}S)$, if $\Sigma\beta_i$ is odd, and let $N = M(0;\frac{1}{2}S,(1,1))$, if $\Sigma\beta_i$ is even. Then $H_1(N;\mathbb{F}_2) = 0$, as can be seen by reducing the entries of the matrix in Theorem *mod* (2). Hence N is the 2-fold branched cover of a knot, by the observations preceding the theorem, and so N_o embeds in S^4 as a fibre of the 2-twist spin $\tau_2 k$.

We may assume that N_o is the complement of a small ball around a point on a regular fibre f, and that $D(\frac{1}{2}S)$ is the complement of a tubular neighbourhood of f. Then $N_o = D(\frac{1}{2}S) \cup_A D^2 \times I$, where A is an annulus in $\partial D(\frac{1}{2}S)$, and N_o has a product neighbourhood $N_o \times [-\epsilon, \epsilon]$ in the knot exterior $X(\tau_2 k)$. This neighbourhood may be written as a union $D(\frac{1}{2}S \times [-\epsilon, \epsilon] \cup (D^2 \times I \times [-\epsilon, \epsilon])$, and so we obtain a smooth embedding of $M(0; S) \cong \partial(D(\frac{1}{2}S) \times [-\epsilon, \epsilon])$ in S^4. Furthermore, $B(M) = D^2 \times I \times [-\epsilon, \epsilon]$ is a ball embedded in S^4 such that $M \cap B(M) = \partial D^2 \times I \times [-\epsilon, \epsilon]$, and each circle $\partial D^2 \times (r, s)$ is a regular fibre of M. □

Issa and McCoy observed that the above result holds also if $\alpha_{2i-1} = \alpha_{2i}$ is even, for one value of i. This is clear if there are just two exceptional fibres, since $M(0; (\alpha, \beta), (\alpha, -\beta)) \cong S^2 \times S^1$, which embeds in S^4.

ADDENDUM. *Let S be skew-symmetric Seifert data with one pair of cone points having even order. Then $M(0; S)$ embeds smoothly in S^4.*

PROOF. If $N = M(0; \frac{1}{2}S)$ with S as above then $H_1(N; \mathbb{F}_2) = 0$, and so the argument of the theorem applies. □

We may use expansion to get embeddings with more such pairs.

LEMMA 3.7. [**IM20**, Lemma 1.6] *If M embeds in S^4 and M_1 is obtained from M by expansion then M_1 embeds in S^4.*

PROOF. Suppose that $M_1 = M \natural_f M'$, where $M' = M(0; (\alpha_1, \beta_1), (\alpha_1, -\beta_1))$ with $(\alpha_1, \beta_1) \in S$ and relatively prime $0 < \beta < \alpha$. Let N_1 be a Seifert fibred neighbourhood of the first exceptional fibre of M, and let $U = N_1 \times [-\frac{1}{2}, \frac{1}{2}]$. Then $\partial U \cong S^2 \times S^1$ has a natural Seifert structure, so that $\partial U = M'$. The restrictions of the fibration to $N_1 \times \{\pm\frac{1}{2}\}$ are copies of the fibration of N_1, and is the obvious product structure on $\partial N_1 \times [-\frac{1}{2}, \frac{1}{2}] \cong T \times [-\frac{1}{2}, \frac{1}{2}]$. Now let $N_2 \subset N_1$ be a Seifert fibred neighbourhood of a regular fibre. Let

$$V = (\overline{M \setminus N_2}) \times \{0\} \cup \partial N_2 \times [-1, -\frac{1}{2}] \cup (\overline{\partial U \setminus N_2} \times \{\frac{1}{2}\}).$$

Then $V \cong M \#_f U \cong M \#_f M'$. By smoothing the corners we obtain a smooth embedding of V into $M \times [-1, 1]$. Thus if M embeds in S^4 so does M_1. □

In Appendix A we show that if ℓ_M is even then all even cone point orders must have the same 2-adic valuation (Lemma A6) and give criteria in terms of the Seifert data for ℓ_M to be hyperbolic (Theorem A7). This suffices to show that $M(0; (4, 1), (4, -1), (2, 1), (2, -1))$ does not embed. On the other hand, if S is skew-symmetric and the even cone point orders are all 2 (or are all 4) then we may use the addendum and Lemma 3.7 to show that $M(0; S)$ embeds. But these cases with small denominator are exceptional. If $M = M(0; (8, 1), (8, -1), (8, 3), (8, -3))$ then ℓ_M is not hyperbolic, and so M does not embed. We do not know whether skew symmetry and hyperbolicity together are sufficient.

We show next that we may use fibre sums to embed $M(g;S)$ when $M(0;S)$ embeds and $g>0$. Given embeddings of two manifolds M and M' in S^4, each with the property stated in Theorem 3.6, we may construct an embedding of $M\sharp_f M'$, also with this property, by taking connected sums of the 4-spheres. Remove the balls $\frac{1}{2}B(M)=D^2\times I\times[-\epsilon,\epsilon]$ and $\frac{1}{2}B(M')$ from the corresponding spheres. The sum is then the union of the closures of the remaining components (with embedded pieces), in the obvious way. Thus, by induction, we obtain a smooth embedding of $M_1\sharp_f\dots\sharp_f M_n$.

LEMMA 3.8. *If $M=M(k;S)$ embeds smoothly into a 4-manifold W, then so does $M\sharp_f M(g;\emptyset)$.*

PROOF. Let f be a regular fibre of M smoothly embedded in W, and let D be a 3-disc with equatorial circle $C\subset\partial D$. Then f has a neighbourhood $D\times f$ in W which is consistently fibred with M so that $N=M\cap(D\times f)$ is a neighbourhood of f in M with boundary $\partial N=C\times f$. Write $W^*=W\setminus int(D\times f)$ and $M^*=M\setminus int(N)$. Then we have a smooth embedding $\phi_M:M^*\to W^*$ such that the restriction $\partial M^*\to\partial W^*=\partial D\times f$ is a fibre-preserving map with image $C\times f$. On the other hand, write $T_g^*=T_g\setminus int(D^2)$. Since there are smooth embeddings of T_g^* in D with $\mathrm{Im}(\partial T_g^*)=C$, there is a smooth fibre-preserving embedding $\phi_T:T_g^*\times f\to D\times f$ with $\mathrm{Im}(\partial T_g^*\times f)=C\times f$. Together ϕ_M and ϕ_T give a smooth embedding of $M\sharp_f M(g;\emptyset)$ into $W=W^*\cup(D\times f)$. □

See [**IM20**, Proposition 7.2] for an alternative account (ascribed to Donald), which uses Kirby calculus to show that $M(g+1;S)$ embeds in $M(g;S)\times[-1,1]$.

It would be convenient to have a converse to this stabilization result. However, when the base is non-orientable there is no such cancellation. See Chapter 4.

Together Theorem 3.6 and Lemma 3.8 imply the following.

COROLLARY 3.8.1. *Let S be skew-symmetric Seifert data for which all cone point orders α_i are odd. Then $M(g;S)$ embeds smoothly in S^4 for all $g\geqslant 0$.* □

3.5. $\mathbb{H}^2\times\mathbb{E}^1$-Manifolds with $g=0$

When the Seifert data is skew-symmetric, all cone point orders are odd and $\varepsilon(M)=0$ then M embeds smoothly in S^4, by Theorem 3.6. We shall show that if $g=0$, $\varepsilon(M)=0$ and all cone point orders are odd, then skew symmetry is a necessary condition for $M(0;S)$ to embed in a homology 4-sphere. If $r\leqslant 2$ then M must be $S^2\times S^1$, while if $r\geqslant 4$ then M is a $\mathbb{H}^2\times\mathbb{E}^1$-manifold.

If the S^1-action extends to a fixed-point free action onto a complementary region W then there is a direct, geometric argument. The exceptional orbits with non-trivial isotropy subgroup have even codimension and are foliated by circles. Therefore, they are tori (in the interior of the region) and annuli with boundary components exceptional fibres of M, and $\chi(W)=0$. Consideration of relative orientations implies that the Seifert data of the boundary components of such an annulus have the form $\{(\alpha,\beta),(\alpha,-\beta)\}$.

THEOREM 3.9. *Let $M = M(g; S)$, and let $\phi : \pi_1 M \to \mathbb{Z}$ be an epimorphism such that $\phi(h) \neq 0$. If b_ϕ is neutral and $\alpha_i > 2$ for all i then S is skew-symmetric.*

PROOF. We note first that $\varepsilon(M) = 0$ since the image of h in $H_1(M)$ has infinite order. Let $\sigma = |\phi(h)|$. Since h is central, $\phi^{-1}(\sigma\mathbb{Z}) \cong \mathrm{Ker}(\phi) \times \mathbb{Z}$, and so the covering space associated to this subgroup is a product $F \times S^1$, where F is a closed surface. Then M fibres over S^1 with fibre F and monodromy θ of order σ, and so the base orbifold B is the quotient of F by an effective action of $G = \mathbb{Z}/\sigma\mathbb{Z}$.

Let $\varphi : \pi^{orb}(B) = \pi_1 M/\langle h \rangle \to \mathbb{Z}/\sigma\mathbb{Z}$ be the epimorphism induced by ϕ, and let $g \in G$ have image $\varphi(g) = [1]$. Then F is the covering space associated to $\mathrm{Ker}(\varphi)$. The points P with non-trivial isotropy subgroup G_P lie above the cone points of B, and the representation of the isotropy subgroup $G_P = \langle g^{\sigma/\alpha_i} \rangle$ on the tangent space T_P determines, and is determined by, the Seifert invariant (α_i, β_i), corresponding to the ith cone point of B, since $\varphi(q_i) = -[\frac{\sigma}{\alpha}{}_i\beta_i]$, for all $i \leqslant r$.

Since $M_\phi \cong F \times \mathbb{R}$ the Blanchfield pairing b_ϕ on $H_1(M_\phi; \mathbb{Q})$ reduces to the intersection pairing I_F on the fibre F, together with the isometric action of $G = Aut(F/B) \cong \mathbb{Z}/\sigma\mathbb{Z}$. Let $s(n,k) = |\{j : \alpha_j = n, \beta_j \equiv k \ mod\ (n)\}|$ and $K(m) = \{k : 1 \leqslant k < m/2, (k,m) = 1\}$. Then the equivariant signatures of the G-Signature Theorem are given by

$$sign(I_F, g^{\sigma/m}) = i\Sigma_{k \in K(m)}(s(m, m-k) - s(m,k)) \cot(\frac{\pi k}{m}).$$

[**AB68**, Theorem 6.27]. (See also [**Go86**] and Section 1.9.)

If b_ϕ is neutral, its image in the Witt group $W_+(\mathbb{Q}(t), \mathbb{Q}\Lambda)$ is trivial. The equivariant signature $sign(I_F, g^{\sigma/m})$ is an invariant of the Witt class of b_ϕ, for each $m|\sigma$. (See page 75 of [**Neu**].) Since b_ϕ is neutral these signatures are 0. If $m > 2$, the algebraic numbers $\{\cot(\frac{\pi k}{m}) : k \in K(m)\}$ are linearly independent over $\mathbb{Q}$ [**EE82**]. Hence $s(m, m-k) = s(m,k)$ for all $k \in K(m)$ and $m > 2$. Since we may modify each β_i by multiples of α_i, subject to $\varepsilon_S = 0$, the Seifert data is skew-symmetric. □

The converse is also true, but we shall only sketch the argument, as we do not need the result. Suppose that $S = ((\alpha_1, \beta_1), \dots, (\alpha_{2s}, \beta_{2s}))$ is skew-symmetric, and let $T = \{(\alpha_{2j}, \beta_{2j}) \mid j \leqslant s\}$. Then $M \cong N \#_f - N \#_f M(g; \emptyset)$, where $N = M(0; T)$, and ϕ induces non-zero homomorphisms $\phi_N : \pi_1 N \to \mathbb{Z}$ and $\psi : \pi_1 M(g; \emptyset) \to \mathbb{Z}$. Let $n(f)$ be an open tubular neighbourhood of a regular fibre of N, and let $N_0 = N \setminus n(f)$. Then $N \#_f - N = N_0 \cup_\partial -N_0$. If $g = 0$, then ϕ_N is an epimorphism, and the diagonal copy of $H_1(N_0; \mathbb{Q}\Lambda)$ in $H_1(M; \mathbb{Q}\Lambda) \cong H_1(N_0; \mathbb{Q}\Lambda) \oplus H_1(N_0; \mathbb{Q}\Lambda)$ is a maximal self-annihilating submodule, so b_ϕ is neutral. In general, ϕ_N and ψ need not be onto, and we must allow for infinite cyclic covering spaces with finitely many components.

The argument for the following theorem extends that of [**BB22**, Theorem 2.4].

THEOREM 3.10. *Let W be a compact 4-manifold with connected boundary M, and let $\pi = \pi_1 W$. Suppose that $H^1(W) \cong H^1(M) \cong \mathbb{Z}$ and that $H_2(W) = 0$. Let $\phi : \pi \to \mathbb{Z}$ be an epimorphism representing a generator of $H^1(W)$. Then the Blanchfield pairing on $H_1(M_\phi; \mathbb{Q}) = H_1(M; \mathbb{Q}\Lambda)$ is neutral.*

PROOF. Consideration of the Wang sequences associated to the infinite cyclic covers of M, W and (W, M) shows that $t-1$ acts epimorphically on $H_1(M;\mathbb{Q}\Lambda)$, $H_1(W;\mathbb{Q}\Lambda)$ and $H_2(W;\mathbb{Q}\Lambda)$. Since $\mathbb{Q}\Lambda$ is noetherian and W and M are compact these modules are finitely generated, and so are $\mathbb{Q}\Lambda$-torsion modules on which $t-1$ acts invertibly. Poincaré duality then implies that the same is true for all the homology and cohomology modules of M, W and (W, M) with coefficients $\mathbb{Q}\Lambda$.

Let j be the inclusion of M as the boundary of W, and let $\delta : H_2(W, M;\mathbb{Q}\Lambda) \to H = H_1(M;\mathbb{Q}\Lambda)$ be the connecting homomorphism in the long exact sequence of homology. Let $P = \mathrm{Im}(\delta)$ be the image of $H_2(W, M;\mathbb{Q}\Lambda)$ in H. Poincaré duality combined with the Universal Coefficient Theorem gives us isomorphisms of the three short exact sequences:

$$\begin{array}{ccccc}
0 \to P & \longrightarrow & H = H_1(M;\mathbb{Q}\Lambda) & \longrightarrow & \mathrm{Im}(j_*) \cong H/P \to 0 \\
\downarrow{\scriptstyle PD} & & \downarrow{\scriptstyle PD} & & \downarrow{\scriptstyle PD} \\
0 \to \overline{\mathrm{Im}(j^*)} & \longrightarrow & \overline{H^2(M;\mathbb{Q}\Lambda)} & \longrightarrow & \overline{\mathrm{Im}(\delta^*)} \to 0 \\
\downarrow{\scriptstyle UCT} & & \downarrow{\scriptstyle UCT} & & \downarrow{\scriptstyle UCT} \\
0 \to \overline{\mathrm{Im}(\mathrm{Ext}(j_*))} & \longrightarrow & \overline{\mathrm{Ext}_{\mathbb{Q}\Lambda}(H, \mathbb{Q}\Lambda)} & \longrightarrow & \overline{\mathrm{Im}(\mathrm{Ext}(\delta))} \to 0.
\end{array}$$

Hence $\Delta_0(H/P) = \overline{\Delta_0(P)}$. This is enough to show that the "Alexander polynomial" $\Delta_0(H)$ is a product $\lambda\overline{\lambda}$, where $\lambda = \Delta_0(P)$.

The central isomorphism from H to $\overline{\mathrm{Ext}_{\mathbb{Q}\Lambda}(H, \mathbb{Q}\Lambda)} = \mathrm{Hom}(H, \mathbb{Q}(t)/\mathbb{Q}\Lambda)$ is the adjoint $\tilde{b}_\phi$ of the Blanchfield pairing. Let

$$P^\perp = \{u \in H \mid b_\phi(u, p) = 0 \ \forall p \in P\}.$$

Then we have another commutative diagram

$$\begin{array}{ccccc}
0 \to P^\perp & \longrightarrow & H & \longrightarrow & H/P^\perp \to 0 \\
\downarrow & & \downarrow{\scriptstyle \tilde{b}_\phi} & & \downarrow \\
0 \to \overline{\mathrm{Ext}_{\mathbb{Q}\Lambda}(H/P, \mathbb{Q}\Lambda)} & \longrightarrow & \overline{\mathrm{Ext}_{\mathbb{Q}\Lambda}(H, \mathbb{Q}\Lambda)} & \longrightarrow & \overline{\mathrm{Ext}_{\mathbb{Q}\Lambda}(P, \mathbb{Q}\Lambda)} \to 0.
\end{array}$$

Hence $\Delta_0(H/P^\perp) = \overline{\Delta_0(P)}$. Comparing these equations, and using the multiplicity of orders in short exact sequences, we see that $\Delta_0(P^\perp) = \Delta_0(P)$.

If we use the geometric formulation of the Blanchfield pairing as in the proof of [**AIL**, Theorem 2.4], we see easily that $P \leqslant P^\perp$. Since $\mathbb{Q}\Lambda$ is a PID and these are torsion modules of the same order it follows that $P = P^\perp$. Hence b_ϕ is neutral. □

COROLLARY 3.10.1. *Let $M = M(0; S)$, where $\varepsilon(M) = 0$ and all cone point orders α_i are odd. Then M embeds in S^4 if and only if S is skew-symmetric.*

PROOF. In this case $\beta = 1$, and so $H^1(X) \cong H^1(M) \cong \mathbb{Z}$. Let $\phi : \pi_X \to \mathbb{Z}$ be an epimorphism. The restriction $\phi : \pi = \pi_1 M \to \mathbb{Z}$ is also an epimorphism, since $H_1(M)$ maps onto $H_1(X)$, and $H_1(M;\mathbb{Q}(t)) = 0$. Thus b_ϕ is neutral, by Theorem 3.10, and so S is skew-symmetric, by Theorem 3.9. □

This settles the question of which Seifert fibred rational homology $S^2 \times S^1$s embed, when all cone point orders are odd; the answer is not known in general.

3.6. An Open Question

We consider some hindrances in extending our argument to the cases with cone points of even order. Skew symmetry alone is not enough to ensure an embedding, since the 2-torsion must be homogeneous, by Lemma A6 of Appendix A.

If there is an embedding with $H_2(X) = 0$ then the inclusions of M and $\vee^\beta S^1$ into X induce isomorphisms on all the rational lower central series quotients of the fundamental groups [**St65**]. Hence these quotients are those of the free group $F(\beta)$. This is never true for M an $\mathbb{H}^2 \times \mathbb{E}^1$-manifold with $\beta > 1$, since the centre has non-zero image in $H_1(M; \mathbb{Q})$. It follows immediately from Lemma 2.2 that if $g \leqslant 1$ then there is a complementary region X with $\chi(X) = 0$. Does h have non-zero image in $H_1(X; \mathbb{Q})$ when $\chi(X) = 0$ and $g = 1$?

If $g > 1$, the condition $\chi(X) = 0$ holds for neither complementary region of $M(g; \emptyset) = T_g \times S^1$, when embedded in S^4 as the boundary of a regular neighbourhood of an embedding of T_g. It remains possible that b_ϕ is neutral when $\Phi(h) \neq 0$ for some Φ as above. (This would follow by the argument of [**AIL**, Theorem 2.3], if the torsion submodule of $H_2(X, M; \mathbb{Q}\Lambda)$ maps onto the image of $H_2(X, M; \mathbb{Q}\Lambda)$ in $H_1(M; \mathbb{Q}\Lambda)$.)

When $r = 2$ and $\varepsilon(M) = 0$ the Seifert data S is skew-symmetric, and $M = M(0; S) \cong S^2 \times S^1$, with S^1-action given by $u.(w, z) = (u^\alpha w, u^\beta z)$ for $u, z \in S^1$ and $w \in \widehat{\mathbb{C}} \cong S^2$. If a regular fibre of M bounds a locally flat disc in one complementary component of some embedding of M in S^4, ambient surgery gives an embedding of $L\# -L$ in S^4, where $L = L(\alpha, \beta)$. This is only possible if α is odd [**KK80**]. It is not clear whether a fibre sum construction can be used to build up embeddings of other $\mathbb{H}^2 \times \mathbb{E}^1$-manifolds with some exceptional fibres of even multiplicity.

If $M = M(0; (3, 1), (5, -2), (15, 1))$, then the cone point orders α_i are odd, $\varepsilon_S = 0$ and $\tau_M = 0$, so ℓ_M is hyperbolic. Thus these conditions alone do not imply that S must be skew-symmetric.

Closed orientable graph manifolds may be characterized as the 3-manifolds whose JSJ decompositions have no hyperbolic pieces. They form the largest class of 3-manifolds with a good natural parametrization (in terms of weighted graphs). Although we shall not consider general graph manifolds in this book, we might expect that the question of which orientable graph manifolds embed in S^4 would admit an answer in terms of their parametrizations. A closed orientable 3-manifold may have distinct representations as a graph manifold. The paper [**Neu81**] is the definitive account of the parametrization of graph manifolds and of the "calculus" of transformations relating different representations of such manifolds.

CHAPTER 4

Seifert Manifolds with Non-orientable Base Orbifolds

In this chapter we shall consider Seifert manifolds with non-orientable base orbifold.

W. S. Massey settled an old question of Whitney when he showed that the normal bundles to smooth embeddings of $\#^k\mathbb{RP}^2$ in S^4 have Euler number $e \in \{-2k, 4-2k, \ldots, 2k\}$ [**Ma69**]. This was an early application of the G-Signature Theorem to low-dimensional topology. We shall use this theorem in a similar manner in proving the main result of this chapter, which gives a bound on the range of values for $\varepsilon(M)$, where M embeds in S^4.

4.1. The Torsion Subgroup

We begin by determining the torsion τ_M for Seifert manifolds M with non-orientable base orbifold.

Let N_i be a regular neighbourhood of the ith exceptional fibre, and let B_o be a section of the restriction of the Seifert fibration to $\overline{M \setminus \cup\{N_i\}}$. Then B_o is homeomorphic to $\#^k RP^2$ with r open 2-discs deleted, and $\overline{M \setminus \cup\{N_i\}} \cong B_o \times S^1$. Let ξ_i and θ_i be simple closed curves on ∂N_i corresponding to the ith boundary component of B_o and a regular fibre respectively. Since B is non-orientable $\pi_1 M$ has a presentation

$$\langle u_1, \ldots, u_k, q_1, \ldots, q_n, h \mid \Pi u_i^2 \Pi q_j = 1,\ q_i^{\alpha_i} h^{\beta_i} = 1,\ u_j h u_j^{-1} = h^{-1} \rangle,$$

where the generators u_i represent orientation-reversing loops in the $\mathbb{RP}^2$ summands of $|B|$, q_j is the image of ξ_j and h is the image of regular fibres such as θ_j. In this case $u_j h u_j^{-1} = h^{-1}$ for all j, since M is orientable.

LEMMA 4.1. *Let $M = M(-k; S)$ be a Seifert manifold with non-orientable base orbifold (so $k \geqslant 1$). Let r be the maximal index such that $t_i = t_1$ for $i \leqslant r$. Let $c = r$ if $t_1 \neq 0$ and let $c = \Sigma\beta_i$ if $t_1 = 0$ (i.e., if all cone point orders α_i are odd). Then*

$$\tau_M \cong \mathbb{Z}/2\alpha_1\mathbb{Z} \oplus \mathbb{Z}/2\alpha_2\mathbb{Z} \oplus \Big(\bigoplus_{i=3}^{n} \mathbb{Z}/\alpha_i\mathbb{Z}\Big) \quad \text{if } c \text{ is even},$$

and

$$\tau_M \cong \mathbb{Z}/4\alpha_1\mathbb{Z} \oplus \Big(\bigoplus_{i=2}^{n} \mathbb{Z}/\alpha_i\mathbb{Z}\Big) \quad \text{if } c \text{ is odd}.$$

PROOF. On abelianizing the above presentation for $\pi_1 M$, we see that $H_1(M)$ has the presentation

$$\langle a_1, \ldots, a_k, q_1, \ldots, q_n, h \mid 2h = 0,\ \alpha_j q_j + \beta_j h = 0,\ 2\Sigma a_i + \Sigma q_j = 0\rangle,$$

where $h = [f]$, $a_i = [u_i]$ and $q_j = [\xi_j]$, for $1 \leqslant i \leqslant k$ and $1 \leqslant j \leqslant n$. Since h, q and $a = \Sigma a_i$ are clearly torsion elements, we see that $H_1(M) \cong \mathbb{Z}^{k-1} \oplus \tau_M$, where τ_M has the presentation

$$\langle a, h, q_1 \ldots, q_n \mid 2h = 0,\ \alpha_j q_j + \beta_j h = 0,\ 2a + \Sigma q_j = 0\rangle.$$

We shall write the cone point orders as $\alpha_i = 2^{t_i} s_i$, with s_i odd, and in descending order of 2-adic valuation t_i.

Suppose first that $t_1 > 0$. Then after modifying the Seifert data, if necessary, we may assume that every β_i is odd, and so rewrite the appropriate relations in the above presentation for τ_M as $\alpha_j q_j = h$. Replace q_j by $\bar{q}_j = q_j + 2^{-t_j}\alpha_1 q_1$, and replace a with

$$\bar{a} = a - \left(\Sigma_{j=2}^{n} \frac{\alpha_1}{2^{t_j+1}}\right) q_1 + \Sigma_{j=3}^{n} \frac{1}{2}(1 - s_j)\bar{Q}_j + \omega,$$

where $\omega = \frac{1}{2} q_1$ if r is even and $\omega = \frac{1}{2}(1 - s_2)\bar{q}_2$ if r is odd. It is easily checked that the coefficient of q_1 in this expression is an integer, in either case. Now the presentation for τ_M reduces to one of the following:

$$\langle h, \bar{a}, q_1, \bar{q}_2, \ldots, \bar{q}_n \mid 2h = 0,\ \alpha_1 q_1 = h,\ \alpha_j \bar{q}_j = 0,\ 2\bar{a} + \bar{q}_2 + \Sigma_{j=3}^{n} s_j \bar{q}_j = 0\rangle,$$

or

$$\langle h, \bar{a}, q_1, \bar{q}_2, \ldots, \bar{q}_n \mid 2h = 0,\ \alpha_1 q_1 = h,\ \alpha_j \bar{q}_j = 0,\ 2\bar{a} + \bar{q}_1 + \Sigma_{j=2}^{n} s_j \bar{q}_j = 0\rangle,$$

according as r is even or odd.

If r is even, replace $\bar{q}_2$ by $\bar{\bar{q}}_2 = \bar{q}_2 + \Sigma_{j=3}^{n} s_j \bar{q}_j$. With this, the relation $\alpha_2 \bar{q}_2$ may be replaced by $\alpha_2 \bar{\bar{q}}_2 = 0$ (noting that α_j divides $\alpha_2 s_j$ for $j \geqslant 3$), and the presentation for τ_M becomes

$$\langle h, \bar{a}, q_1, \bar{\bar{q}}_2, \bar{q}_3, \ldots, \bar{q}_n \mid 2h = 0,\ \alpha_1 q_1 = h,\ \alpha_2 \bar{\bar{q}}_2 = \alpha_j \bar{q}_j = 0,\ 2\bar{a} + \bar{\bar{q}}_2 = 0\rangle,$$

which reduces to

$$\langle \bar{a}, q_1, \bar{q}_3, \ldots, \bar{q}_n \mid 2\alpha_1 q_1 = 0,\ 2\alpha_2 \bar{a} = 0,\ \alpha_j \bar{q}_j = 0\rangle,$$

and so $\tau_M \cong \mathbb{Z}/2\alpha_1\mathbb{Z} \oplus \mathbb{Z}/2\alpha_2\mathbb{Z} \oplus (\bigoplus_{i=3}^{n} \mathbb{Z}/\alpha_i\mathbb{Z})$.

If r is odd, replace q_1 by $\bar{q}_1 = q_1 + \Sigma_{j=2}^{n} s_j \bar{q}_j$. With this, the relation $\alpha_1 q_1 = h$ may be replaced by $\alpha_1 \bar{q}_1 = h$, and the presentation for τ_M becomes

$$\langle h, \bar{a}, \bar{q}_1, \ldots, \bar{q}_n \mid 2h = 0,\ \alpha_1 \bar{q}_1 = h,\ \alpha_j \bar{q}_j = 0\ (j \geqslant 2),\ 2\bar{a} + \bar{q}_1 = 0\rangle,$$

which reduces to

$$\langle \bar{a}, \bar{q}_1, \ldots, \bar{q}_n \mid 4\alpha_1 \bar{q}_1 = 0,\ \alpha_j \bar{q}_j = 0\ (j \geqslant 2)\rangle,$$

and so $\tau_M \cong \mathbb{Z}/4\alpha_1\mathbb{Z} \oplus (\bigoplus_{i=2}^{n} \mathbb{Z}/\alpha_i\mathbb{Z})$.

Finally, suppose that $t_1 = 0$ and every α_j is odd. After reindexing, if necessary, we may assume that β_j is odd for $j \leqslant s$ and is even for $j > s$, for some $s \geqslant 0$. Then $\alpha_j q_j = h$ for $j \leqslant s$ and $\alpha_j q_j = 0$ for $j > s$, and $s \equiv c = \Sigma_{j=1}^n \beta_j \; mod\ (2)$. Let

$$\bar{q}_j = q_j + \alpha_1 q_1 = q_j + h \quad \text{for } 2 \leqslant j \leqslant s \quad \text{and} \quad \bar{q}_j = q_j \quad \text{for } j > s,$$

and let

$$\bar{a} = a - \left(\Sigma_{j=2}^s \frac{1}{2}\alpha_1\right) q_1 + \Sigma_{j=3}^n \frac{1}{2}(1 - s_j)\bar{Q}_j + \omega,$$

where $\omega = \frac{1}{2}q_1$ if c is even and $\omega = \frac{1}{2}(1 - s_2)\bar{q}_2$ if c is odd. Then the rest of the argument follows exactly as in the preceding cases. □

If M is as in the above lemma then τ_M is a direct double only if the prime power factors of the cone point orders α_i occur in pairs. If α_i is odd for all i then we need also that $\Sigma_{i=1}^n \beta_i$ is even. In particular, this requires that there are an even number $2b$ (possibly 0) of cone points with α_i even. As we shall see in Theorem 4.3, hyperbolicity of the torsion linking form requires furthermore that the non-zero t_is must all be equal, or, when all the cone point orders are odd, that $\Sigma_{i=1}^n \alpha_i \beta_i \equiv 2k$ *mod* (4).

If M is Seifert fibred over a nonorientable base then $\tau_M \neq 0$, by Lemma 4.1, and so M is neither a homology sphere nor a homology $S^2 \times S^1$.

4.2. The Linking Pairing

In order to determine the linking pairing we shall have to look more carefully at geometric representatives for the homology classes.

We begin by finding a convenient choice of generators for τ_M. Fix a base-point for $\#^k\mathbb{RP}^2$ and let $\sigma_1, \ldots, \sigma_k$ be orientation-reversing loops corresponding to the distinct summands. Choose loops $\xi_1, \ldots, \xi_n$ at $*$ which bound discs N_i, which meet only at $*$. Orient and label these loops (clockwise) as in Figure 4.1.

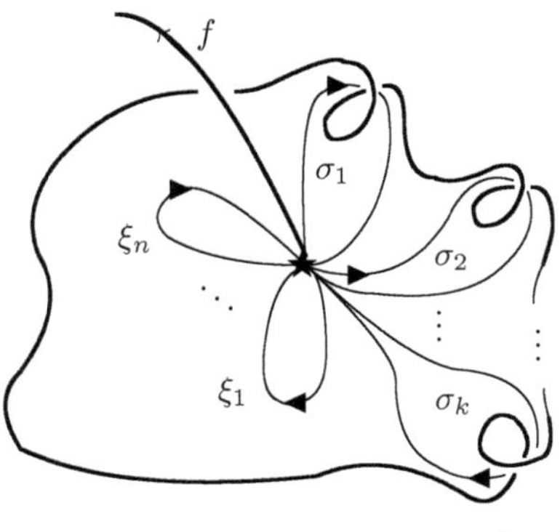

Figure 4.1 Some curves on $\#^k\mathbb{RP}^2$.

Let $P_{k,n*} = \#^k\mathbb{RP}^2 \setminus \cup\, intN_i$. Then $P_{k,n*}$ deformation retracts onto a wedge of circles and thus there is an unique orientable S^1-bundle $P_{k,n*}\tilde{\times}S^1$ over $P_{k,n*}$, with a section which carries the loops σ_i and ξ_j into $P_{k,n*}\tilde{\times}S^1$. Let f denote the oriented fibre over $* \in P_{k,n*}$. Finally let $N(\gamma_i)$ be solid tori with cores γ_i, for $1 \leqslant i \leqslant n$, and write

$$M = P_{k,n*}\tilde{\times}S^1 \cup_{\cup h_i} (\bigcup_{i=1}^{n} N(\gamma_i)),$$

where each attaching homomorphism h_i identifies the boundary of a meridianal disc in $N(\gamma_i)$ with a curve representing the homology class $\alpha_i[\xi_i] + \beta_i[f]$. Then $M = M(-k; S)$, where $S = \{(\alpha_i, \beta_i) \mid 1 \leqslant i \leqslant n\}$, and the cores γ_i represent the singular fibres.

Now consider the loops σ_i, ξ_j and f of Figure 4.1 to be 1-cycles in M. Let $\sigma = \Sigma_{i=1}^{k}\sigma_i$. Then σ has image $a = [\sigma]$ in τ_M. The relations in our original presentation for τ_M correspond to the presence of the following 2-chains in M. The union of fibres over σ_1 may be considered as a 2-chain A (with support the image of an annulus), with $\partial A = 2f$. (Recall that each σ_i is an orientation-reversing loop!) For each $i \leqslant n$ there is a 2-chain D_i in $N(\gamma_i)$ with $\partial D_i = \alpha_i\xi_i + \beta_i f$ in M. (Thus D_i is homologous to a meridianal disc.) Finally, we may consider $P_{k,n*}$ as a 2-chain in M with $\partial P_{k,n*} = 2\sigma + \Sigma_{i=1}^{n}\xi_i$. Note that we have described each of ∂A, $\partial P_{k,n*}$ and the ∂D_is explicitly as 1-cycles. We may now compute various linkings.

LEMMA 4.2. *Using the above generators for τ_M, we have*

$$\ell_M(q_i, q_i) = \frac{\beta_i}{\alpha_i}, \quad \ell_M(q_i, q_j) = 0 \text{ if } j \neq i, \quad \ell_M(q_i, h) = \ell_M(h, h) = 0,$$

$$\ell_M(a, q_i) = -\frac{\beta_i}{2\alpha_i}, \quad \ell_M(a, a) = \frac{1}{4}(2k - \varepsilon(M)) \quad \text{and} \quad \ell_M(a, h) = \frac{1}{2}.$$

PROOF. The 1-cycles corresponding to the generators are shown in Figure 4.1, with f directed out of the page, and we may use the right-hand rule to determine a positive intersection. We may perturb each ξ_i to a homologous 1-cycle ξ_i' just inside $N(\gamma_i)$, and perturb σ to σ' as illustrated in Figure 4.2.

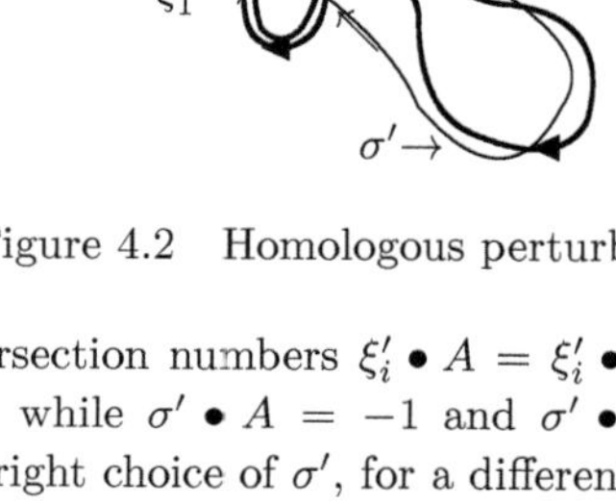

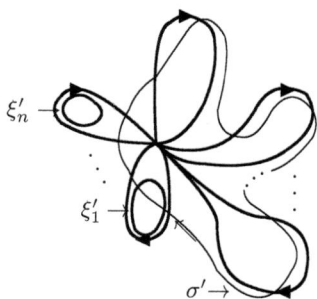

Figure 4.2 Homologous perturbations.

Then we have intersection numbers $\xi_i' \bullet A = \xi_i' \bullet P_{n*} = 0$, $\xi_i' \bullet D_i = \beta_i$ and $\xi_i' \bullet D_j = 0$ if $j \neq i$, while $\sigma' \bullet A = -1$ and $\sigma' \bullet D_i = -\beta_i$. We also have $\sigma' \bullet P_{n*} = k$ (with the right choice of σ', for a different choice may give $-k$). Now let $C_j = 2D_j - \beta_j A$, so that $\partial C_j = 2\alpha_j\xi_j$. Then $\xi_i' \bullet C_j = 2\xi_i' \bullet D_j = \delta_{ij}2\beta_i$, and consequently $\ell_M(q_i, q_j) = \delta_{ij}\frac{\beta_i}{\alpha_i}$. Also $\sigma' \bullet C_i = -2\beta_i + \beta_i = -\beta_i$, so that $\ell_M(a, q_i) = -\frac{\beta_i}{2\alpha_i}$. Now write $C = \Sigma_{i=1}^n \widehat{\alpha}_i C_i$, where $\widehat{\alpha}_i = \alpha_i^{-1}\Pi_{j=1}^n\alpha_j$, and let $R = 2(\Pi_{j=1}^n\alpha_j)P_{n*} - C$. Then $\partial R = 4(\Pi_{j=1}^n\alpha_j)\sigma$. Also

$$\sigma' \bullet R = 2(\Pi_{j=1}^n\alpha_j)k + \Sigma_{i=1}^n\widehat{\alpha}_i\beta_i = (\Pi_{j=1}^n\alpha_j)(2k - \varepsilon(M)).$$

Thus $\ell_M(a, a) = \frac{1}{4}(2k - \varepsilon(M)$. (Note that we may also use R to check that $\ell_M(q_i, a) = \frac{-\beta_i}{2\alpha_i} = \ell_M(a, q_i)$, which also follows by the symmetry of the pairing.) Finally, $\xi_i' \bullet A = 0$ implies that $\ell_M(q_i, h) = 0$ for $i \leqslant n$, and $\sigma' \bullet A = -1$ implies that $\ell_M(a, h) = -\frac{1}{2} = \frac{1}{2}$, since $\partial A = 2f$. Also we may perturb f to f' so that $f' \bullet A = 0$. Thus $\ell_M(h, h) = 0$. □

COROLLARY 4.2.1. *If* $M = M(-1; ((2, 1), (2, -1), (1, -e))$ *then* ℓ_M *is hyperbolic if and only if* $e \equiv 2$ *mod* (4).

PROOF. In this case $\tau_M \cong (\mathbb{Z}/4\mathbb{Z})^2$, with generators a and q_1, and the matrix of ℓ_M with respect to this basis is

$$L = \begin{pmatrix} \frac{1}{4}(2-e) & -\frac{1}{4} \\ -\frac{1}{4} & \frac{1}{2} \end{pmatrix}.$$

It is easily checked that there is a matrix $A \in GL(2, \mathbb{Z}/4\mathbb{Z})$ such that $A^{tr}LA = \begin{pmatrix} 0 & 1 \\ 1 & 0 \end{pmatrix}$ if and only if $e \equiv 2$ mod (4). □

In particular, the Hantzsche-Wendt flat manifold $M(-1; (2, 1), (2, -1))$ does not embed in any homology 4-sphere. On the other hand, 0-framed surgery on the link 8_2^2 gives an embedding of the $\mathbb{N}il^3$-manifold $M(-1; (2, 1), (2, -1), (1, 2))$.

THEOREM 4.3. *Let* $M = M(-k; S)$. *If* ℓ_M *is hyperbolic then all the even cone point orders* α_i *must have the same* 2*-adic valuation. If all cone point orders are odd then we must also have* $\eta = \Sigma\alpha_i\beta_i \equiv 2k$ *mod* (4).

Remark. The quantity η is invariant *mod* (4) under Seifert data equivalences.

PROOF. Enlarge the set of generators to $\{c_1, \dots, c_n, d_1, \dots, d_n, h, \hat{a}\}$, where $c_i = 2^{t_i}q_i - \beta_i h$ and $d_i = s_i q_i$ for $i \leqslant n$, and $\hat{a} = a + \Sigma_{i=1}^n \frac{1}{2}\rho_i(s_i + 1)c_i$, with ρ_i chosen so that $2^{t_i}\rho_i \equiv 1$ *mod* (s_i). This corresponds to a decomposition of $H_1(M)$ into a direct sum of n cyclic factors $\langle c_i \rangle$ of odd order s_i, for $i \leqslant n$, and a 2-group. By using Lemma 4.2, one can check that, with respect to this set of generators, the linking matrix for ℓ_M is a block sum of the diagonal matrix

$$\ell_M(c_i, c_j) = \operatorname{diag}\left(\frac{2^{t_1}\beta_1}{s_1}, \dots, \frac{2^{t_n}\beta_n}{s_n}\right)$$

with the 2-group block

$$\begin{pmatrix} \frac{s_1\beta_1}{2^{t_1}} & 0 & \dots & 0 & -\frac{\beta_1}{2^{t_1+1}} \\ 0 & \frac{s_2\beta_2}{2^{t_2}} & \dots & 0 & -\frac{\beta_2}{2^{t_2+1}} \\ \vdots & \vdots & \ddots & \vdots & \vdots \\ 0 & 0 & \dots & 0 & \frac{1}{2} \\ -\frac{\beta_1}{2^{t_1+1}} & -\frac{\beta_2}{2^{2_1+1}} & \dots & \frac{1}{2} & x \end{pmatrix},$$

where $x = \ell_M(\hat{a}, \hat{a})$. When every α_i is odd this block simplifies to $\begin{pmatrix} 0 & \frac{1}{2} \\ \frac{1}{2} & \frac{1}{4}(2k+\eta) \end{pmatrix}$, which is itself hyperbolic if and only if $2k + \eta \equiv 0 \ mod \ (4)$, which gives the second part of the theorem.

Now suppose that, contrary to the conclusion of the first part of the theorem, there is some $m \in \{2, \ldots, n\}$ such that $0 < t_m < t_1 = t$. It remains to show in this case that ℓ_M is not hyperbolic. Note that α_1 and α_m are even, so that β_1 and β_m are odd (as well as s_1 and s_m being odd). Replace the generator d_m with $\bar{d}_m = d_m + 2^{t-t_m} s_m d_1$ (or just $\bar{d}_m = s_m \bar{q}_m$ for $\bar{q}_m$ as in the proof of Lemma 4.1. Then $\ell_M(\bar{d}_m, h) = 0$, $\ell_M(\bar{d}_m, d_i) = 0$ for $i \neq 1$ or m and

$$\ell_M(\bar{d}_m, \bar{d}_m) = \ell_M(d_m, d_m) + (2^{t-t_M} s_m)^2 \ell_M(d_1, d_1) = \frac{u}{2^{t_m}},$$

where $u = s_m \beta_m + 2^{t-t_m} s_m^2 t_1 \beta_1$ is odd, since $t_m < t$ and $s_m \beta_m$ is odd. Moreover

$$\ell_m(\bar{d}_m, d_1) = \frac{v}{2^{t_m}}$$

where $v = s_m s_1 \beta_1$, and

$$\ell_M(\bar{d}_m, \hat{a}) = -\frac{\beta_m}{2^{t_m+1}} - 2^{t-t_m} s - m\frac{\beta_1}{2^{t+1}} + \frac{w}{2^{t_m}},$$

where $w = -\frac{1}{2}(\beta_m + s_m \beta_1)$ is also an integer. Now choosing r such that $ru \equiv 1$ $mod \ (2^{t_m})$, replace d_1 and $\hat{a}$ with $\bar{d}_1 = d_1 - vr\bar{d}_m$ and $\bar{a} = \hat{a} - wr\bar{d}_m$, respectively. Then $\ell_M(\bar{d}_m, \bar{d}_1) = 0$ and $\ell_M(\bar{d}_m, \bar{a}) = 0$. Thus, $\bar{d}_m$ generates an orthogonal direct summand of the 2-group. But a hyperbolic linking on a 2-group admits no such summand [**KK80**, Theorem 0.1]. □

Using Wall's presentation of the semigroup of linkings on p-groups (for a fixed odd prime p) [**KK80, Wa64**], it is a routine matter to check whether or not the odd-order part of a given torsion linking pairing is hyperbolic. In the case of Proposition 4.4.1, where the strictly singular fibres occur in opposite pairs, we can easily check that $\eta \equiv -\varepsilon \ mod \ (4)$. Moreover, since

$$\begin{pmatrix} \frac{\beta_i}{\alpha_i} & 0 \\ 0 & -\frac{\beta_i}{\alpha_i} \end{pmatrix} \sim \begin{pmatrix} 0 & \frac{1}{\alpha_i} \\ \frac{1}{\alpha_i} & 0 \end{pmatrix},$$

the linking pairing is clearly hyperbolic in this case if and only if $\varepsilon \equiv 2k \ mod \ (4)$. Manifolds with all cone point orders α_i even may be treated case by case with the help of the presentation of the semigroup of linkings on 2-groups given in [**KK80**].

4.3. The Main Result

We shall assume that $\alpha_i = 2^{t_i} s_i$ with s_i odd, for all i. We may also assume that there is an $r \geqslant 0$ such that $t_1 = \cdots = t_r$, and $t_i = 0$ if $i > r$, by Theorem 4.3. In what follows we shall write $t = t_1$. Since ℓ_M must be hyperbolic there is an even number of cone points with α_i even. Let $2b$ be this number, let $b_S = b - 1$ if $b > 0$ and let $b_S = 0$ if all the cone point orders are odd.

THEOREM 4.4. *Let S be strict Seifert data. If $M_1 = M(-k_1; S, (1, e_1))$ and $M_2 = M(-k_2; S, (1, e_2))$ each embed in homology 4-spheres then*

$$(\varepsilon(M_1) - \varepsilon(M_2)) \in \mathbb{Z} \cap 2^{-t}\{-2K, 4-2K, \ldots, 2K-4, 2K\},$$

where $K = 2^t(k_1 + k_2 + 4b_S) + b_S$.

In particular, if the cone point orders α_i are all odd (so $t = b_S = 0$) and $M(-k; S, (1, e))$ embeds in a homology 4-sphere then there are at most k other values of e for which the corresponding 3-manifold embeds in a homology 4-sphere.

COROLLARY 4.4.1. *Suppose that* $S = \{\gamma_i(\alpha_i, \beta_i), \gamma_i(\alpha_i, \alpha_i - \beta_i) \mid 1 \leqslant i \leqslant n\}$, *where each* α_i *is odd, and let* $M = M(-c; S, (1, e))$. *Then* $\varepsilon(M) = -n - e$, *and* M *embeds in a homology* 4*-sphere if and only if* $-2c \leqslant \varepsilon(M) \leqslant 2c$ *and* $\varepsilon(M) \equiv 2c$ mod (4). *Every such manifold embeds smoothly in* S^4.

PROOF. On the one hand, there are at most $c + 1$ possible values of e corresponding to 3-manifolds which embed in homology 4-spheres, by Theorem 4.4.

On the other hand, we may adapt the construction of Theorem 3.6. There we saw that $M(S^2; ((\alpha_i, \beta_i), (\alpha_i, -\beta_i) \cong \partial(L_i^* \times [-\epsilon, \epsilon])$ has a smooth embedding in S^4 for which there exists a 4-ball $B(M_i) = D^2 \times I \times [-\epsilon, \epsilon]$ embedded in S^4 such that $M_i \cap B(M_i) = \partial D^2 \times I \times [-\epsilon, \epsilon]$, and each circle $\partial D^2 \times (r, s)$ is a regular fibre of M_i. The embeddings of $M(-c; (1, e))$ (for e in the above range) due to Massey [**Ma69**] also have this property, since in fact each one bounds a disc bundle. As in Theorem 3.6 we may form embedded fibre sums to obtain the desired embedding. These must exhaust the possibilities. □

Of course we may change the ambient homology 4-sphere by taking connected sums.

COROLLARY 4.4.2. *Let* M *be an* S^1 *bundle over a closed surface* F, *with Euler number* e. *Then* M *embeds in a homology* 4*-sphere if and only if either*

(1) F *is orientable and* $e = 0$ *or* ± 1; *or*

(2) $F = \#^c\mathbb{RP}^2$ *and* $e \in \{-2c, 4 - 2c, \dots, 2c\}$.

In all cases M *embeds smoothly in* S^4.

PROOF. The case when F is non-orientable follows immediately from Corollary 4.4.1. If F is orientable of genus g then $H_1(M) \cong \mathbb{Z}^{2g} \oplus (\mathbb{Z}/e\mathbb{Z})$, and so τ_M is a direct double if and only if $e = 0$ or ± 1. Since $M(g; (1, 0)) = M(g; \emptyset) \cong T_g \times S^1$. while $M(0; (1, \pm 1)) \cong S^3$ and $M(g; (1, \pm 1)) \cong T_{g-1} \times S^1$ for $g > 0$, these all embed smoothly. □

Thus orientable S^1-bundle spaces over non-orientable bases embed if and only if they embed smoothly, as boundaries of regular neighbourhoods of smooth embeddings of the base. Note also that although $M(-3; (1, 6)) = M(-1; (1, 6)\#_f T \times S^1$ embeds in S^4, we cannot cancel the $T \times S^1$ term as $M(-1; (1, 6))$ does not embed.

LEMMA 4.5. *Let* $M = M(-k; S)$. *Then one of the complementary regions* U *has a* 2^{t+1}*-fold covering* $\psi : \widetilde{U} \to U$ *such that the restriction of the covering involution to* $\widetilde{M}$ *acts by rotating the fibres through a half-turn. (Here we assume that* $\widetilde{M}$ *has the Seifert structure induced by the covering* $\psi|_M$.)

PROOF. Since τ_M must be a direct double, we may assume by Lemma 4.1 that it has a direct summand $\mathbb{Z}/2\alpha_1\mathbb{Z}$ generated by q_1, as in the proof. In fact $\widehat{q} = s_1 q_1$

generates a direct summand of τ_M of order 2^{t+1}. Note also that $h = \alpha_1 q_1 = 2^t \widehat{q}$, where h is the element of order 2 in $H_1(M)$ which is the image of a regular fibre f of M. Define an epimorphism $\phi : H_1(M) \to \mathbb{Z}/2^{t+1}\mathbb{Z}$ by projection onto the summand of τ_M generated by $\widehat{q}$. Let V be the other complementary region. Now $\widehat{q}$ may be identified with an element $(u_0, v_0) \in H_1(U) \oplus H_1(V)$, and one of $(u_0, 0)$ or $(0, v_0)$ must have order 2^{t+1} under the quotient map ϕ. Suppose that u_0 satisfies this condition. Thus ϕ defines a surjection $\phi_X : H_1(U) \to \mathbb{Z}/2^{t+1}\mathbb{Z}$ such that $u \mapsto \phi(u, 0)$, and there exists a 2^{t+1}-fold covering $\psi : \widetilde{U} \to U$ with $\pi_1 \widetilde{U} = \operatorname{Ker}(\phi_U \circ \mathrm{ab})$.

Since the inclusion $M \subset U$ induces the projection $p_U : H_1(M) \to H_1(U)$, the regular fibre f of M represents the element $[f] = p_U(h) = 2^t u_0$ in $H_1(U)$, and since $\phi_U(u_0)$ has order 2^{t+1}, $\phi_U([f])$ will have order 2 in $\mathbb{Z}/2^{t+1}\mathbb{Z}$. Thus $\psi|_M^{-1}(f)$ is a collection of 2^t disjoint circles, each of which doubly covers f. On the other hand, if we take

$$\overline{\phi}_U : H_1(X) \to \mathbb{Z}/2^{t+1}\mathbb{Z} \to \mathbb{Z}/2^t\mathbb{Z}$$

such that $[f] \mapsto 0$, then we obtain an intermediate covering $\overline{\psi}_U : \overline{U} \to U$, where each fibre of M lifts to a fibre in $\overline{M}$. Thus $\widetilde{U}$ is a double cover of $\overline{U}$ with covering automorphism as required. That is, each regular fibre of $\widetilde{M}$ doubly covers a fibre of $\overline{M}$, the same being necessarily true for singular fibres as well. □

LEMMA 4.6. *Let M be as in the previous lemma. Then $\widetilde{M}$ is fibred over $\#^c\mathbb{RP}^2$, where $c = 2^t(k + b_S) - 2b_S$ and $\varepsilon(\widetilde{M}) = 2^{t-1}\varepsilon(M)$.*

PROOF. Clearly $\widetilde{M}$ and $\overline{M}$ each have the same base orbifold $\widetilde{B}$. Observe also that the cone point orders of $\widetilde{B}$ must all be odd since, given a fibre γ in $\overline{M}$ of type (α, β), β must be even as γ is covered by a fibre $\tilde{\gamma}$ of type $(\alpha, \frac{1}{2}\beta + c\alpha)$ [**NR78**, 1.3]. Now, in the covering $\overline{M} \to M$ each regular fibre of M lifts to 2^t distinct fibres, as do the singular fibres of odd multiplicity, while each of the remaining fibres, having multiplicity $2^t s_i$ for some odd s_i, must be 2^t-fold covered by a single fibre of odd multiplicity. The covering $\widetilde{M} \to \overline{M}$ induces a covering of orbifolds, or a 2^t-fold branched covering of surfaces $|\widetilde{B}| \to |B|$ with a branch set consisting of the $2b$ points corresponding to the $2b$ singular fibres with even multiplicity. Thus $|\widetilde{B}|$ is also a non-orientable surface, and so $|\widetilde{B}| \cong \sharp^c\mathbb{RP}^2$, for some $c > 0$. Recall that, if $t \neq 0$ then there are $2b_S + 2$ fibres of even multiplicity, by definition of b_S. In this case $\chi(|\widetilde{B}|) = 2^t(\chi(|B|) - 2b_S - 2) + 2b_S + 2$, which gives $c = 2^t(k + 2b_S) - 2b_s$. But this formula also holds when $t = 0$, in which case $c = k$. Finally, $\varepsilon(\widetilde{M}) = \frac{1}{2}\varepsilon(\overline{M}) = 2^{t-1}\varepsilon(M)$, as in [**NR78**, 1.2] or [**JN**]. □

We aim to construct involutions g on a 4-manifold W such that g has fixed point set a closed connected surface F which is smoothly embedded in W, and g acts smoothly on a neighbourhood of F. That is, F will have a smooth D^2-bundle neighbourhood $N(F) \subset W$, on which g acts by rotation of the fibres. We shall call such a surface a "good" fixed point set. Given such an F, let $e(F)$ be the normal Euler number, which is just the Euler number of $\partial N(F)$, considered as a circle bundle over F. According to the G-Signature Theorem, if g is an involution of a closed connected 4-manifold with "good" fixed point set F then

$\text{sign}(g, W) = e(F)$ [**AS68, Go86**]. In proving Theorem 4.4, we shall use also the fact that $|\text{sign}(g, W)| \leqslant \beta_2(W)$, and is congruent to $\beta_2(W)$ *mod* (2), for W a closed 4-manifold. (See also Section 1.9.)

If we take M to be the circle bundle $M(F; (1, -e))$ then, in Lemma 4.5, $\widetilde{M}$ is also a circle bundle, with Euler number $\tilde{e} = \frac{1}{2}e$ and associated disc bundle $D_{\widetilde{M}}$. In this case the covering involution of $\widetilde{X}$ given by Lemma 4.5, extends to an involution g of the closed connected 4-manifold $W_{\widetilde{M}} = \widetilde{X} \cup_{\widetilde{M}} D_{\widetilde{M}}$, with "good" fixed point set $\widetilde{F} \cong F$. Thus $\text{sign}(g, W) = e(\widetilde{F}) = \tilde{e}$.

More generally, given a Seifert manifold $\widetilde{M}$ with fibration $\tilde{\pi} : \widetilde{M} \to \widetilde{F}$, the "associated disc bundle", or more precisely the mapping cylinder

$$C_{\widetilde{M}} = \widetilde{M} \times [0, 1]/(x_1, 0) \sim (x_2, 0) \text{ whenever } \tilde{\pi}(x_1) = \tilde{\pi}(x_2),$$

is in fact not necessarily a manifold. It ceases to be a manifold precisely at the isolated points $p_i = (x_i, 0)$ for x_i on a strictly singular fibre. However, we may modify $C_{\widetilde{M}}$ by removing a neighbourhood which contains all such points. There exists a closed disc $\delta \subset \widetilde{F}$ such that the point $\tilde{\pi}(\tilde{\gamma}) \in \text{int}(\delta)$ for each singular fibre $\tilde{\gamma}$ of $\widetilde{M}$. Let $E = \tilde{\pi}^{-1}(\delta) \times [0, \frac{1}{2})/(\sim) \subset C_{\widetilde{M}}$. This defines a neighbourhood which meets $\widetilde{F}$ in δ. (Note that $C_{\widetilde{M}}$ contains $\widetilde{F}$ as $\{(x, 0) \in C_{\widetilde{M}}\}$.) Define the "modified associated disc-bundle" to be $D_{\widetilde{M}} = C_{\widetilde{M}} \setminus \text{int}(E)$.

Note that in the above construction E is contractible to $\delta \sim *$, so that $\chi(E) = 1$. Also its boundary ∂E must be a rational homology sphere. This is true since ∂E is closed, connected and orientable, and has finite first homology, since it may be written as a generalized Seifert manifold $M(0; (0, 1), (\tilde{\alpha}_1, \tilde{\beta}_1), \ldots, (\tilde{\alpha}_m, \tilde{\beta}_m))$.

As in the statement of the theorem, for fixed S, let $M_1 = M(k_1, S, (1, e_1))$ and $M_2 = M(k_2, S, (1, e_2))$ be Seifert manifolds which embed in homology 4-spheres Σ_1 and Σ_2, respectively. Then we may write $\Sigma_1 = X_1 \cup_{M_1} Y_1$ and $\Sigma_2 = X_2 \cup_{M_2} Y_2$, where we may suppose there are coverings $\psi_1 : \widetilde{X}_1 \to X_1$ and $\psi_2 : \widetilde{X}_2 \to X_2$, as described in Lemma 4.5.

Consider the manifold $W_1 = \widetilde{X}_1 \cup_{\widetilde{M}_1} D_{\widetilde{M}_1}$, with boundary ∂E_1. By Lemma 4.5, there is a covering involution of $\widetilde{X}_1$ which extends to an involution g_1 of W_1 in the obvious way, with fixed point set $\widetilde{F}_1 \setminus \text{int}(\delta_1)$. There is a similar pair (g_2, W_2). Since $\widetilde{M}_1$ and $\widetilde{M}_2$ have the same set $\widetilde{S}$ of strictly singular fibres, we may write $\widetilde{M}_i = M(-\tilde{k}_i; (\tilde{\alpha}_1, \tilde{\beta}_1), \ldots, (\tilde{\alpha}_m, \tilde{\beta}_m), (1, -e_i))$, and we may choose E_1 and E_2 to have exactly the same fibred structure. That is, there is a fibre-preserving homeomorphism $E_1 \cong E_2$ which respects the sets $\widetilde{F}_i \cap E_i$ and the involution on ∂E_i. Now, if we write

$$W = W_1 \cup_{\partial E} W_2,$$

then W is a closed connected 4-manifold, and g_1 and g_2 together define an involution g on W with "good" fixed point set $F = \widetilde{F}_1 \sharp \widetilde{F}_2$. Furthermore F has normal Euler number $e(F) = e_1 - e_2 = \varepsilon(\widetilde{M}_1) - \varepsilon(\widetilde{M}_2)$, which is simply $2^{t-1}(\varepsilon(M_1) - \varepsilon(M_2))$, by Lemma 4.6. Note that $\partial(-W_2) \cong -\partial E$, while $\partial(W_1) \cong \partial E$. However, in order to orient the union of two oriented manifolds consistently, we require the attaching homeomorphism to be orientation-reversing, and here it is taken to be the obvious one. Now, by the G-Index Theorem we may write

$$\operatorname{sign}(g,W) = 2^{t-1}(\varepsilon(M_1) - \varepsilon(M_2)).$$

This signature is bounded by the value of $\beta_2(W)$, independently of the values of $\varepsilon(M_1)$ and $\varepsilon(M_2)$. It remains now to express this constraint in terms of k_1 and k_2, and the constants t and b_S, which depend only on S. We begin by establishing a bound on $\beta_1(W_i)$ (for $i = 1, 2$), for which we need the next lemma.

LEMMA 4.7. *Let p be a prime. If $f : \widetilde{X} \to X$ is a p^s-fold cyclic covering then $\beta_1(\widetilde{X}) \leqslant p^s \beta_1(X; \mathbb{F}_p)$.*

PROOF. The covering gives an exact sequence

$$1 \to K_0 \to G_0 \to \mathbb{Z}/p^s\mathbb{Z} \to 1,$$

where $K_0 = \pi_1 \widetilde{X}$ and $G_0 = \pi_1 X$. Since K_0 is finitely generated, $\overline{K} = K_0/I(K_0)$ is free abelian of rank $r = \beta_1(\widetilde{X})$. Since $I(K_0)$ is characteristic in K_0 it is normal in G_0, and so we have an exact sequence

$$1 \to \overline{K} \to G_0/I(K_0) \to \mathbb{Z}/p^s\mathbb{Z} \to 1.$$

Let $K = \overline{K}/p\overline{K} \cong (\mathbb{Z}/p\mathbb{Z})^r$. Then we get another exact sequence

$$1 \to K \to G \to \mathbb{Z}/p^s\mathbb{Z} \to 1,$$

where G is the quotient of $G_0/I(K_0)$ by $p\overline{K}$. Note that G^{ab} is a quotient of $H_1(X; \mathbb{Z})$.

Now $G \cong \langle \zeta, K \mid \zeta^{p^s} \in K,\ \zeta^{-1}x\zeta = Ax,\ \forall x \in K \rangle$, where ζ acts on K via some $A \in GL(r,p)$. Let $\Gamma = \mathbb{F}_p[T]$. Then K is a Γ-module, with T acting via $t.x = Ax$, for all $x \in K$. Since $\zeta^{p^s} \in K$ and K is abelian, $T^{p^s}x = x$, for all $x \in K$, and so K is annihilated by $T^{p^s} - I = (T - I)^{p^s}$. Since Γ is a PID we have $K \cong \oplus_{i=1}^m (\Gamma/(T-I)^i)^{e_i}$, for some $e_i \geqslant 0$ and $m \leqslant p^s$. Hence $r = \Sigma_{i=1}^m i e_i \leqslant p^s \Sigma e_i$. On the other hand, $G^{ab} = \langle \zeta, K/(T-I)K \mid \zeta^{p^s} \in K/(T-I)K \rangle$, and $K/(T-I)K \cong \oplus_{i=1}^m (\Gamma/(T-I))^{e_i} \cong \mathbb{F}_p^R$, where $R = \Sigma e_i$. Thus $\beta_1(\widetilde{X}) \leqslant p^s R$ and $R \leqslant \beta_1(X; \mathbb{F}_p)$. □

LEMMA 4.8. $\beta_1(W_i) \leqslant \beta_1(\widetilde{X}_i) \leqslant 2^{t+1}(\beta_1(X_i) + b_S)$.

PROOF. For convenience, we shall drop the subscript i throughout, except to distinguish W_i from W. Apply Lemma 4.7 to the covering ψ of Lemma 4.5, where $p^s = 2^{t+1}$. The image of ζ in G^{ab} has order 2^{t+1} or 2^{t+2}. But τ_X has no elements of order 2^{t+2}. Hence ζ has order 2^{t+1} and $G^{ab} \cong \mathbb{Z}/2^{t+1}\mathbb{Z} \oplus (\mathbb{Z}/2\mathbb{Z})^R$, where R is as defined in Lemma 4.7. Since $\tau_M \cong \tau_X \oplus \tau_Y$ is a direct double, $\tau_X/2\tau_X$ has dimension $b_S + 1$, by Lemma 4.1. Hence we get an epimorphism from $(\mathbb{Z}/2\mathbb{Z})^{\beta_1(X)+b_S+1}$ to $(\mathbb{Z}/2\mathbb{Z})^{R+1}$, and hence $R \leqslant \beta_1(X) + b_S$. But then $\beta_1(\widetilde{X}) \leqslant 2^{t+1}(\beta_1(X) + b_S)$.

Write $W^0 = W_i \cup_{\partial E} E = \widetilde{X} \cup_{\widetilde{M}} C_{\widetilde{M}}$. (Recall that $C_{\widetilde{M}}$ is contractible to $\widetilde{F}$.) Then we have the Mayer-Vietoris sequence

$$H_1(\widetilde{M}) \xrightarrow{\ \phi\ } H_1(\widetilde{X}) \oplus H_1(\widetilde{F}) \to H_1(W^0) \to 0,$$

where ϕ is induced by the inclusions of $\widetilde{M}$ into $\widetilde{X}$ and $\widetilde{C}_{\widetilde{M}}$. But the map to $H_1(\widetilde{F})$ is surjective, so $\dim(\operatorname{Im}(\phi)) \geqslant \beta_1(F)$, and hence $\beta_1(W^0) = \beta_1(\widetilde{X}) + \beta_1(\widetilde{F}) - \dim(\operatorname{Im}(\phi)) \leqslant \beta_1(\widetilde{X})$. Finally, since $\beta_1(E) = 0$ and ∂E is a rational homology sphere, the Mayer-Vietoris sequence (for $W^0 = W_i \cup_{\partial E} E$) gives $\beta_1(W_i) = \beta_1(W^0)$, and we have proved the lemma. □

We may now prove Theorem 4.4.

PROOF. For each $i = 1, 2$ the union $C_{\widetilde{M}_i} = D_{\widetilde{M}_i} \cup E_i$ gives $\chi(C_{\widetilde{M}_i}) = \chi(D_{\widetilde{M}_i}) + \chi(E_i)$. That is, $\chi(D_{\widetilde{M}_i}) = \chi(\widetilde{F}_i) - 1$, since $C_{\widetilde{M}_i}$ is contractible to $\widetilde{F}_i$ and $\chi(E_i) = 1$. Thu, since $\Sigma = \widetilde{X}_1 \cup D_{\widetilde{M}_1} \cup D_{\widetilde{M}_2} \cup \widetilde{X}_2$, where the unions are along closed 3-manifolds, we have

$$\chi(W) = 2^{t+1}\chi(X_1) + 2^{t+1}\chi(X_2) + \chi(\widetilde{F}_1) + \chi(\widetilde{F}_2) - 2.$$

By Poincaré duality, we may then write

$$\begin{aligned}\beta_2(W) &= 2(\beta_1(W_1) + \beta_1(W_2) + 2^{t+1}(\chi(X_1) + \chi(X_2)) + \chi(\widetilde{F}_1) + \chi(\widetilde{F}_2) - 4 \\ &= \Xi_1 + \Xi_2, \quad \text{where} \quad \Xi_i = 2\beta_1(W_i) + 2^{t+1}\chi(X_i) + \chi(\widetilde{F}_i) - 2,\end{aligned}$$

where the Mayer-Vietoris sequence (for $W = W_1 \cup_{\partial E} W_2$) gives $\beta_1(W) = \beta_1(W_1) + \beta_1(W_2)$, since ∂E is a rational homology sphere.

Since X_i is a complementary region for an embedding of M_i in a homology 4-sphere Σ_i, $\beta_1(X_i) + \beta_2(X_i) = \beta_1(M_i) = k_i - 1$, while $\beta_j(X_i) = 0$ for $j > 2$. The inequality of Lemma 4.8 and these relations now give

$$\begin{aligned}\Xi_i &\leqslant 2^{t+2}(\beta_1(X_i) + b_S) + 2^{t+1}\chi(X_i) + \chi(\widetilde{F}_i) - 2 \\ &= 2^{t+1}(1 + \beta_1(X_i) + \beta_2(X_i) + 2b_S) + \chi(\widetilde{F}_i) - 2 \\ &= 2^{t+1}(k_i + 2b_S) - \tilde{k}_i = K_i, \quad \text{say.}\end{aligned}$$

Consequently, putting $K = K_1 + K_2$, we have $|\text{sign}(g, W)| \leqslant \beta_2(W) \leqslant K$. Also $\Xi_i \equiv \chi(\widetilde{F}_i) \equiv \tilde{k}_i \equiv K_i \ mod\ (2)$, so that $\text{sign}(g, W) \equiv \beta_2(W) \equiv K \ mod\ (2)$. Thus $2^{t-1}(\varepsilon(M_1) - \varepsilon(M_2)) = \text{sign}(g, W) \in \{-K, 2-K, \ldots, K\}$. Hence, since $\varepsilon(M_1) - \varepsilon(M_2)$ must be an integer, it follows that

$$\varepsilon(M_1) - \varepsilon(M_2) \in \{-\frac{2K}{2^t}, \frac{4-2K}{2^t}, \ldots, \frac{2k}{2^t}\} \cap \mathbb{Z}.$$

Finally, note that since $\tilde{k}_i = 2^t(k_i + 2b_S) - 2b_S$, by Lemma 4.6, we have $K_i = 2^2(k_i + 2b_S) = 2b_S$, and so $K = 2^t(k_1 + k_2 + 4b_S) + 4b_s$, which completes the proof. □

4.4. Some Further Remarks

Let $M = M(-c; S)$ be a Seifert manifold with base $B = \#^k\mathbb{RP}^2(\alpha_1, \ldots, \alpha_r)$, where $c > 0$ and all cone point orders α_i are odd. (Then M is an $\mathbb{H}^2 \times \mathbb{E}^1$-manifold or a $\widetilde{\mathbb{SL}}$-manifold unless $c + r \leqslant 2$.) Since $\varepsilon(M)$ is a rational number with odd denominator it has a well-defined image in $\mathbb{Z}/2^s\mathbb{Z}$, for any $s \geqslant 0$. In particular, $\varepsilon(M) \equiv \Sigma\beta_i \ mod\ (2)$ and $\varepsilon(M) \equiv -\Sigma\alpha_i\beta_i \ mod\ (4)$, since $\alpha_i \equiv 1 \ mod\ (2)$ and $\alpha_i \equiv \alpha_i^{-1} \ mod\ (4)$ if α_i is odd. The invariants c and η used in Lemma 4.1 and Theorem 4.3 are just the images of $-\varepsilon(M)$ in $\mathbb{Z}/2\mathbb{Z}$ and $\mathbb{Z}/4\mathbb{Z}$ (respectively). Hence $\varepsilon(M) \equiv 2k \ mod\ (4)$ if ℓ_M is hyperbolic, by Theorem 4.3.

If $\varepsilon(M) = 0$, then $\tau_M \cong (\mathbb{Z}/2\mathbb{Z})^2 \oplus \bigoplus_{i\geqslant 1} \mathbb{Z}/\alpha_i\mathbb{Z}$, by Theorem 4.3. Therefore, if τ_M is a direct double the numbers $\#\{i : v_p(\alpha_i) = j\}$ are even for all odd primes p and exponents $j \geqslant 1$. If, moreover, $r = 3$, then it follows from Lemma 4.2 that ℓ_M is hyperbolic. Must S be skew-symmetric if $\varepsilon(M) = 0$ and $M(-k, S)$ embeds in S^4? The first cases to test are perhaps those with base $P_2(p, q, pq)$, where p and q are distinct odd primes. (When $k = 2$, the complementary domain X satisfies $H_1(M; \mathbb{Q}) \cong H_1(X; \mathbb{Q})$ and $\chi(X) = 0$. However, the argument of Theorem 3.9 does not appear to apply usefully here, as we must first pass to the 2-fold cover of M induced by the orientation cover of the base B before continuing to an infinite cyclic cover homeomorphic to $F \times \mathbb{R}$. There is no obvious reason that the Blanchfield pairing associated to the latter cover should be neutral.)

Donald shows that if S' is strict Seifert data and $M = M(-c; S', (1, -e))$ embeds smoothly in S^4 then the Seifert invariants must occur in pairs $\{(\alpha, \beta), (\alpha, \beta')\}$ where $\beta' = \alpha - \beta$ or $\beta' = \alpha - \beta^{[-1]}$, with $\beta.\beta^{[-1]} \equiv 1 \ mod\ (\alpha)$. Moreover, if any of the cone point orders are even then all such cone point orders are equal, and if (α, β) and (α, β') are Seifert invariants with α even then $\beta' = \beta$, $\alpha - \beta$, $\beta^{[-1]}$ or $\alpha - \beta^{[-1]}$ [**Do15**, Theorem 1.2].

CHAPTER 5

Smooth Embeddings

Although our main interest is in the topological case, as outlined in the Preface, the results of Donald, Issa and McCoy on smooth embeddings of Seifert 3-manifolds are a natural complement to our material. In this short chapter we shall outline how this work depends on Donaldson's Diagonalization Theorem, and comment briefly on some other aspects of smooth embeddings of 3-manifolds in 4-manifolds. We shall state the major results of [**Do15**] and [**IM20**], but shall not attempt to give full details.

5.1. Constraints on the Intersection Pairing

Arguments based on the intersection pairings of orientable 4-manifolds with given boundary have been prominent. We mention here two outstanding results which distinguish smooth 4-manifolds from other 4-manifolds.

The first such result is Rokhlin's Theorem. Recall that the intersection pairing of a closed orientable 4-manifold Z is *even* if $\alpha \bullet \alpha$ is even for all α in $H_2(Z)$. (It is enough to check this for a basis of $H_2(Z)/(torsion)$.)

THEOREM (Rokhlin). [**Rok52**] *Let Z be a closed orientable smooth 4-manifold such that $w_2(Z) = 0$. Then $\sigma(Z)$ is a multiple of 16.* □

The hypothesis $w_2(Z) = 0$ is satisfied if $H_1(Z)$ has no 2-torsion and the intersection pairing of Z is even. Extensions and refinements of this classic result are treated at length in [**GM**]. The following non-embedding result was an early application of Rokhlin's Theorem.

EXAMPLE 5.1. *S^3/I^* does not embed smoothly.*
The complementary regions of any embedding j of a homology sphere in S^4 are homology 4-balls. The homology sphere $M = S^3/I^*$ bounds a plumbed 4-manifold W with intersection pairing E_8, which is even, positive definite and of rank 8. (See Example 3.1.) If $j : M \to S^4$ were a smooth embedding then $W \cup_M X$ and $W \cup_M Y$ would be closed orientable smooth 4-manifolds with signature 8, which would contradict Rokhlin's theorem.

There is a related invariant. An orientable smooth n-manifold Z is a *Spin*-manifold if $w_2(Z) = 0$. Closed orientable 3-manifolds have trivial tangent bundles, and so are *Spin*-manifolds. Moreover, if $\mathfrak{s}$ is a *Spin*-structure on M then it bounds; there is a compact 4-manifold W with $\partial W \cong M$ and with a *Spin*-structure which restricts to $\mathfrak{s}$ on M. If W' is a second such 4-manifold then $Z = W \cup -W'$ is a closed smooth *Spin*-manifold and $\sigma(Z) = \Sigma(W) - \sigma(W')$, by Novikov additivity.

Hence the image of $\frac{\sigma(W)}{16}$ in $\mathbb{Z}/16\mathbb{Z}$ does not depend on the choice of *Spin*-manifold W. This is the *Rokhlin invariant* $\mu(M,\mathfrak{s})$ of $(M,\mathfrak{s})$. If $H_1(M)$ is finite of odd order then it has only one *Spin* structure, and we may write just $\mu(M)$. In particular, if M is a homology sphere then $\mu(M) \in 8\mathbb{Z}/16\mathbb{Z} = \mathbb{Z}/2\mathbb{Z}$.

The most striking applications thus far of more recent 4-dimensional differential topology for the study of embeddings have used the Diagonalization Theorem of Donaldson.

THEOREM (Donaldson). [**Do87**] *Let Z be a closed, orientable smooth 4-manifold. If the intersection pairing $\bullet$ on $H = H_2(Z)/(torsion)$ is positive definite, then it is diagonalizable: H has a basis $\{h_1, \dots, h_r\}$ such that $h_i \bullet h_i = 1$ for all i and $h_i \bullet h_j = 0$ if $i \neq j$.* □

We shall show next that the plumbed manifolds $W(g;S)$ with boundary $M(g;S)$ have definite intersection pairings if $\varepsilon(S) > 0$. Of course these are not closed 4-manifolds, but we shall correct that in the next section.

Since $W(\Gamma_S)$ has a handle decomposition with one 0-handle, m 2-handles and no 1-, 3- or 4-handles, it is simply connected and $H_2(W(\Gamma_S))$ is torsion free. This group has a natural basis $\{v_1, \dots, v_m\}$ corresponding to the cores of the 2-handles. (We shall assume that v_1 corresponds to the central vertex.) Let Q_{Γ_S} be the intersection form on $H_2(W(\Gamma_S)) \cong \mathbb{Z}^m$. Then $Q_{\Gamma_S}(v_i, v_j) = e_i$ if $i = j$, -1 if the ith and jth vertices are connected by an edge, and 0 otherwise. The *intersection lattice* of $W(\Gamma_S)$ is the pair $(\mathbb{Z}^m, Q_{\Gamma_S})$.

It is easy to see that if $e = Q_{\Gamma_S}(v_1, v_1) \geqslant s$ then Q_{Γ_S} is positive definite, since the other diagonal entries are all $\geqslant 2$ and the non-zero off-diagonal entries are sparse. With a little more care we may improve on this.

LEMMA 5.2. [**NR78**, Theorem 5.2]. *If $\varepsilon(S) > 0$ then Q_{Γ_S} is positive definite, and if $\varepsilon(S) = 0$ then Q_{Γ_S} is semi-definite, with radical $\mathbb{Z}$.*

PROOF. This follows by a double induction, on s and on the length of the ith continued fraction, for each $i \leqslant s$. Let $c^i_j = [a^i_j, a^i_{j+1}, \dots, a^i_{n_i}]_{HJ}$ for all $1 \leqslant i \leqslant s$ and $1 \leqslant j \leqslant n_i$. Then $c^i_j = a_j - \frac{1}{c_{j+1}}$ for $j < n_i$ and $c^i_j > 0$, for all i, j. We shall suppose first that $s = 1$ and simplify the notation by writing just a_j and c_j. We also have $n = m - 1$ and $e_i = a_{i-1}$ if $i \geqslant 1$. Define a new basis $\{v'_1, \dots, v'_m\}$ for $H_2(W(\Gamma_S); \mathbb{Q})$ by a finite downwards induction, as follows. Let $v'_m = v_m$ and $v'_i = v_i + \frac{1}{c_i} v'_{i+1}$ if $i < m$. Then $Q_{\Gamma_S}(v'_i, v'_i) = a_i - \frac{1}{c_{i+1}} = c_i$ for $1 < i \leqslant m$ and $Q_{\Gamma_S}(v'_1, v'_1) = e - \frac{1}{c_1}$, while $Q_{\Gamma_S}(v'_i, v'_j) = 0$, for $i \neq j$. The extension to the cases with $s > 1$ is easy, and so Q_{Γ_S} can be diagonalized (over $\mathbb{Q}$) as $diag[\varepsilon(S), c^1_1, \dots, c^1_{n_1}, c^2_1, \dots, c^s_{n_s}]$. Since $c^i_1 = \frac{\alpha_i}{\beta_i}$ and $c^i_j > 0$ for all i, j, the lemma follows. □

The manifold $W(g;S)$ is obtained from $W(\Gamma_S)$ by deleting open regular neighbourhoods of the dotted components. When $g > 0$ the dotted components have linking number 0 with each of the other components, so the inclusion of $W(g;S)$ into $W(\Gamma_S)$ induces an isomorphism $H_2(W(g;S)) \cong H_2(W(\Gamma_S))$, and the intersection pairing for $W(g;S)$ depends only on S and not on g.

When the base B is non-orientable the central curve does not contribute to H_2, and the intersection form is then determined by the other 2-handles. This is the orthogonal direct sum of intersection forms for plumbed manifolds with boundary a lens space $L(\alpha_i, \beta_i)$, for $1 \leqslant i \leqslant s$, and so is positive definite.

5.2. A Reduction to Linear Algebra

The starting point for the analyses of the smooth embedding problem for Seifert manifolds is that such 3-manifolds bound 4-manifolds with semi-definite intersection forms. We show next that if $\varepsilon(S) > 0$ and $M = M(g; S)$ embeds in S^4 then there are naturally associated closed orientable manifolds with definite intersection pairings. If the embedding is smooth then the Diagonalization Theorem applies to these.

LEMMA 5.3. *Suppose that $g \geqslant 0$ and that $M = M(g; S)$ embeds in S^4, with complementary regions X and Y. Let $W = W(g; S)$. Then at least one of $W \cup_M X$ and $W \cup_M Y$ is positive definite and both are positive definite if $\varepsilon(S) > 0$.*

PROOF. Let $h_\partial : M \to W$ be the inclusion of the boundary, and suppose first that $\varepsilon(S) \neq 0$. Then $H_1(h_\partial; \mathbb{Q})$ is a monomorphism. Since $\beta_2(W) = \beta_2(W, \partial W; \mathbb{Q})$, by Poincaré duality, the natural homomorphism from $H_2(W; \mathbb{Q})$ to $H_2(W, M; \mathbb{Q})$ is also a monomorphism, and so $H_2(h_\partial; \mathbb{Q}) = 0$.

Let $Z = W \cup_M X$. Since $H_1(h_\partial; \mathbb{Q})$ is a monomorphism, $H_2(h_\partial; \mathbb{Q}) = 0$ and $H_2(j_X)$ is an epimorphism, a Mayer-Vietoris argument shows that the inclusion of W into Z induces an isomorphism $H_2(W; \mathbb{Q}) \cong H_2(Z; \mathbb{Q})$. Since W is a codimension-0 submanifold of Z, the intersection number of 2-cycles in W is the same as in Z, and so the rational intersection pairings are isomorphic. Since W is positive definite, so is Z. Similarly for $W \cup_M Y$.

If $\varepsilon(S) = 0$ the kernel K of $H_1(h_\partial)$ is infinite cyclic and so Q_{Γ_S} has radical $\mathbb{Z}$, the image of $H_2(M)$. The kernel K must map injectively to one of $H_1(X)$ or $H_1(Y)$. Hence the inclusion of W into $W \cup_M X$ (or $W \cup_M Y$, respectively) induces an epimorphism, and it induces an isomorphism of rational intersection pairings, modulo the radical. We again conclude that $W \cup_M X$ (or $W \cup_M Y$, respectively) is definite. □

Assume that $\varepsilon(S) > 0$ and that $M = M(g; S)$ embeds smoothly in S^4. Let U be one of the complementary regions and let $Z_U = W(g; S) \cup_M U$. Let $\{h_1, \ldots, h_m\}$ be a basis for $H(W) = H_2(W(g; S))$ and let v_i be the image of h_i in $H(Z_U) = H_2(Z_U)/(torsion)$. Since Z_U is a closed orientable smooth 4-manifold and its intersection pairing is positive definite, by Lemma 5.2, there is an orthonormal basis $\{e_1, \ldots, e_m\}$ for $H(Z_U)$, by the Diagonalization Theorem. Let $A(U)$ be the matrix of the homomorphism from $H(W)$ to $H(Z_U)$ induced by the inclusion. Then $A(U)_{i,j} = e_i \bullet_{Z_U} v_j$, for all i, j, and $Q_{\Gamma_S}(h_i, h_j) = v_i \bullet_{Z_U} v_j = [A(U)^{tr} A(U)]_{i,j}$.

The next result is based on [**GJ11**, Proposition 2.5], and is the link between the Diagonalization Theorem and the detailed combinatorial analyses made by Donald [**Do15**] and Issa and McCoy [**IM20**].

LEMMA 5.4. [**Do15**, Theorem 3.6] *Let $M = M(g; S)$, where $g \geqslant 0$ and $\varepsilon(S) > 0$, and let U be a complementary region of a smooth embedding of M in S^4. Let*

J and K be the images of $H^2(Z)$ and $H^2(W,M)$ in $H^2(W)$. Then $K \leqslant J$ and $J/K \cong \tau_U$.

PROOF. Let $W = W(g;S)$. Then $H_i(W)$ and $H^i(W)$ are torsion-free, for $i \leqslant 2$, and $H^3(W) = 0$. The inclusion of (W,M) into (Z,U) gives rise to a commutative diagram

$$\begin{array}{ccccccc}
H^2(Z,U) & \xrightarrow{\alpha} & H^2(Z) & \xrightarrow{\beta} & H^2(U) & \longrightarrow & H^3(Z,U) \\
\iota_1 \cong \downarrow & & \iota_2 \downarrow & & \iota_3 \downarrow & & \iota_4 \cong \downarrow \\
H^2(W,M) & \xrightarrow{\gamma} & H^2(W) & \xrightarrow{\delta} & H^2(M) & \longrightarrow & H^3(W,M)
\end{array}$$

with exact rows. The homomorphism from $H^2(W,M)/(torsion)$ to $H^2(W)$ induced by γ is a monomorphism with finite cokernel, since the intersection form of W is definite. Hence the image of δ is the torsion subgroup of $H^2(M)$, since $H^3(W,M) \cong H_1(W)$ is torsion-free. This torsion group is isomorphic to τ_M, by the Universal Coefficient Theorem, and so $\operatorname{Im}\delta$ has order $|\tau_U|^2$, by Lemma 2.2.

Let $H^2(W)$ and $H^2(Z_U)/(torsion)$ have bases which are Kronecker dual to the bases $\{h_i\}$ and $\{e_j\}$. The homomorphism induced by ι_2 has matrix $A(U)^{tr}$ with respect to these bases. Let $H^2(W,M)$ have the basis $\{D(h_i)\}$, where $D : H_2(W) \cong H^2(W,M)$ is the Poincaré duality isomorphism. Then the matrix of γ with respect to these bases is Q_{Γ_S}. Since Q_{Γ_S} has matrix $A(U)^{tr}A(U)$ we see that $K \leqslant J$, and that $\operatorname{Im}\delta = \operatorname{Cok}\gamma$ has order $\det A(U)^2$. Hence $|\tau_U| = |\det A(U|$. We also have $J/K \cong \operatorname{Im}\delta\iota_2 = \operatorname{Im} A(U^{tr})/\operatorname{Im} Q_{\Gamma_S}$, and so $|J/K| = |\det A(U| = |\tau_U|$.

The torsion subgroup of $H^2(U)$ is isomorphic to τ_U, by the Universal Coefficient Theorem. A diagram chase shows that this subgroup is in the image of β, since ι_4 is an isomorphism and $H^3(W,M)$ is torsion-free. Hence its image under ι_3 is contained in $\operatorname{Im}\delta\iota_2$, since $\iota_3\beta = \delta\iota_2$. Since ι_3 is a monomorphism, by Lemma 2.2, these subgroups have the same order, and so are equal. Hence $J/K \cong \tau_U$. □

In terms of matrices: $\tau_U \cong \operatorname{Im} A(U)^{tr}/\operatorname{Im} Q_{\Gamma_S}$.

The following corollary is described in [**IM20**] as the " key obstruction" deriving from the Diagonalization Theorem to embedding Seifert manifolds smoothly in S^4.

COROLLARY 5.4.1. [**Do15**, Corollary 3.9] *Let $A(X)$ and $A(Y)$ be the matrices defined as above, and let $H_U = \operatorname{Im} A(U)^{tr}/\operatorname{Im} Q_{\Gamma_S}$, for $U = X$ and Y. Then the $m \times 2m$ matrix $[A(X)^{tr}|A(Y)^{tr}]$ is surjective, and* $\operatorname{Cok} Q_{\Gamma_S} = H_X \oplus H_Y$.

PROOF. Restrictions induce an isomorphism $\tau_M \cong \tau_X \oplus \tau_Y$, by Lemma 2.2. Hence $\operatorname{Cok} Q_{\Gamma_S} \cong H_X \oplus H_Y$, where the direct sum is an internal direct sum of subspaces with trivial intersection. It shall suffice to show that the columns of $[A(X)|A(Y)]$ span $H^2(W)$.

Let $q : H^2(W) \to \operatorname{Cok} Q_{\Gamma_S}$ be the quotient map. If $w \in H^2(W)$ then $q(w) = q(a_X) + q(a_Y)$ for some $a_X \in \operatorname{Im} A(X)^{tr}$ and $a_Y \in \operatorname{Im} A(Y)^{tr}$. Hence $a_X + a_Y = w + k$ for some $k \in \operatorname{Im} Q_{\Gamma_S}$. Since $Q_{\Gamma_S} = A(X)^{tr}A(X)$, $\operatorname{Im} Q_{\Gamma_S} \leqslant \operatorname{Im} A(X)^{tr}$, and so $k \in \operatorname{Im} A(X)^{tr}$. Hence $w = (a_X - k) + (a_Y)$ is in $\operatorname{Im} A(X)^{tr} + \operatorname{Im} A(Y)^{tr}$. □

There are slightly weaker analogues when $g \geqslant 0$ and $\varepsilon(S) = 0$ and when $M = M(-c; S)$ is Seifert fibred over a non-orientable base.

COROLLARY 5.4.2. [**Do15**, Corollary 3.12] *Let $M = M(g; S)$, where $g \geqslant 0$ and $\varepsilon(S) = 0$, and let $m = \beta_2(W(g; S))$. If $M(g; S)$ embeds smoothly in S^4 then Q_{Γ_S} has matrix $A^{tr}A$, where A is an $(m-1) \times m$ matrix of rank $m-1$.* □

COROLLARY 5.4.3. [**Do15**, Corollary 3.13] *Let $M = M(-c; S)$, where $c > 0$, and let $m = \beta_2(W(g; S))$. Suppose that $M(g; S)$ embeds smoothly in S^4, and let $A(U)$ and H_U be defined as above, for $U = X, Y$. Then Q_{Γ_S} has matrix $A(U)^{tr}A(U)$ and $\mathrm{Cok}\, Q_{\Gamma_S} \cong H_U \oplus H_U$, for $U = X, Y$, and $|H_X \cap H_Y| \leqslant 2$.* □

5.3. Seifert Manifolds

We shall now state the main results of Donald [**Do15**] and Issa and McCoy [**IM20**] and shall refer to their papers for details of the arguments. It should be noted that their arguments obstructing smooth embeddings in S^4 also obstruct smooth embeddings in smooth homology 4-spheres.

If the Seifert data for M is skew-symmetric and all the cone point orders are odd then M embeds smoothly in S^4, by Theorem 3.6 and Lemma 3.8. This also holds if there is also one pair $\{(\alpha, \pm\beta)\}$ in S with α even [**IM20**, Proposition 7.8]. However, this is not yet known if more than one pair has even cone point order (except when Theorem 3.7 applies).

Donald has shown that skew symmetry is a necessary condition for a $\mathbb{H}^2 \times \mathbb{E}^1$-manifold to embed smoothly. (The cone point orders are not assumed to be odd here.)

THEOREM. [**Do15**, Theorem 1.3] *Let M be a Seifert manifold with orientable base orbifold and $\varepsilon(M) = 0$. If M embeds smoothly in S^4 then the Seifert data for M is skew-symmetric.* □

The essential input from topology is Corollary 5.4.2. The pursuit of the consequences in [**Do15**] involves a long combinatorial analysis based on the notion of "linear subset" (of a free abelian group with standard definite inner product) introduced by P. Lisca [**Li07**].

Using a closely related argument (involving Corollary 5.4.3 instead of 5.4.2), Donald gives a parallel result for Seifert manifolds with nonorientable base orbifold.

THEOREM. [**Do15**, Theorem 1.2] *Let $M = M(-c; S', (1, -e))$ where $c > 0$ and S' is strict Seifert data. If M embeds smoothly in S^4 then the invariants in S' must occur in pairs $\{(\alpha, \beta), (\alpha, \beta')\}$ where $\beta' = \alpha - \beta$ or $\beta' = \alpha - \beta^{[-1]}$, with $\beta.\beta^{[-1]} \equiv 1$ mod (α). Moreover, if (α, β) and (α', β') are in S' and α and α' are both even then $\alpha' = \alpha$ and $\beta' = \beta$, $\alpha - \beta$, $\beta^{[-1]}$ or $\alpha - \beta^{[-1]}$.* □

Issa and McCoy give strong constraints on which $\widetilde{\mathbb{SL}}$-manifolds with orientable base orbifold embed smoothly in S^4. Assume henceforth that $M = M(g; S', (1, -e)$, where S' is strict Seifert data with index set $\{1, \dots, k\}$ and $e > \Sigma_{i=1}^{k} \frac{\beta_i}{\alpha_i}$, so that $\varepsilon(M) > 0$. (We have used our notational conventions for the Seifert data in what follows.) Their main technical result involves the following notion.

The Seifert manifold M is *partitionable* if τ_M is a direct double and the index set $\{1,\dots,k\}$ has two partitions $\mathcal{P}$ and $\mathcal{P}'$ into e classes $\{C_1,\dots,C_e\}$ such that

(1) $\Sigma_{i\in C_1}\frac{\beta_i}{\alpha_i}=1-\frac{1}{\mathrm{lcm}(a_1,\dots,a_k)}$;
(2) $\Sigma_{i\in C_j}\frac{\beta_i}{\alpha_i}=1$ for $j>1$; and
(3) no non-empty union of a proper subset of classes in $\mathcal{P}'$ is a union of classes in $\mathcal{P}$.

THEOREM. [**IM20**, Theorem 1.4] *If M embeds smoothly in S^4 then M is partitionable.* □

Conditions (1) and (2) of the definition of partitionable use the facts that τ_M is a direct double and that Z_X and Z_Y are smooth manifolds with positive definite intersection pairings, while condition (3) is a consequence of Corollary 5.4.1.

From this they derive the following results (among others).

THEOREM. [**IM20**, Theorem 1.1] *If M embeds smoothly in S^4 then $2e\leqslant k+1$. Moreover, if $k=2e-1$ is odd then M embeds smoothly if and only if*

$$S'=\{e(\alpha,\alpha-1),(e-1)((\alpha,1)\}\quad (\text{and so } \varepsilon(M)=\frac{1}{\alpha}). \qquad \square$$

THEOREM. [**IM20**, Theorem 1.12] *If $k=2e$ and M embeds smoothly in S^4 then there are integers p,q,r,s with $0<q<p$, $0<s<r$, $(p,q)=(r,s)=1$ and $ps+qr+1=pr$ such that S' is either*

(1) $\{(x(p,q),y(r,s),(x-1)(p,p-q),(y-1)(r,r-s)\}$, *where* $x+y=k-2$; *or*
(2) $\{(x(p,q),y(r,s),(x-1)(p,p-q),(y-1)(r,r-s),z(pr,1),z((pr,pr-1)\}$, *where* $x+y+z=k-2$ *and* $z>0$.

Moreover, in case (1) *each such M embeds smoothly in S^4.* □

The argument for the next result uses the Neumann-Siebenmann invariant $\bar{\mu}(M;\mathfrak{s})$, which is an integral lift of the Rokhlin invariant $\mu(M;\mathfrak{s})$.

THEOREM. [**IM20**, Theorem 1.10]. *If $g=0$ and M embeds smoothly in S^4 then $\beta_1(M;\mathbb{F}_2)\leqslant 2e$.* □

They make the following conjectures, and prove the second of these under the additional assumption that all cone point orders are even.

[**IM20**, Conjecture 1.8]. *A Seifert manifold $M(0;S',(1,-e))$ with $\varepsilon(M)>0$ embeds smoothly in S^4 if and only if it is obtained by a (possibly empty) sequence of expansions from some $M(0;S',(1,-1))$ with S' strict Seifert data which also embeds smoothly in S^4.*

[**IM20**, Conjecture 1.9]. *If $M=M(0;S',(1,-e))$ embeds smoothly in S^4 and $\varepsilon(M)>0$ then $k-1\leqslant 2e\leqslant k+1$.*

5.4. Other Work

Much of the other work on smooth embeddings of 3-manifolds in S^4 has concentrated on $\mathbb{Q}$-homology 3-spheres, in particular homology spheres and connected sums of lens spaces.

The results on homology spheres can best be described as sporadic. If a homology sphere embeds in a homology 4-sphere then it bounds an acyclic 4-manifold (i.e., a homology 4-ball). Consideration of embeddings of homology spheres in homology 4-spheres leads naturally to the study of homology cobordism. There are parallel questions about R-homology cobordism of R-homology 3-spheres; the cases $R = \mathbb{F}_2$ and $R = \mathbb{Q}$ are of greatest interest.

Brieskorn homology spheres are the Seifert manifolds $\Sigma(p,q,r) = M(0;S)$, where p,q,r are pairwise relatively prime and $\varepsilon(S) = \frac{1}{pqr}$. Several infinite families of Brieskorn homology spheres are known to embed, but there are also some infinite families which freely generate $\mathbb{Z}^\infty$ subgroups of the homology cobordism group. We refer to the discussions of problems 3.20, 4.2 and 4.5 in [**Kir**] and the surveys [**AMP24**], [**BB22**] and [**Şa24**] for more details and further references.

C. McDonald has revealed a further subtlety of the smooth case: there are infinitely many homology spheres which embed smoothly in homology 4-spheres, but which have no smooth embedding in any smooth homotopy 4-sphere [**Mc22**]. McDonald's argument relies on $SU(2)$-gauge-theoretic results related to definiteness of an intersection pairing.

On the other hand, the long standing question of which connected sums of lens spaces embed smoothly has been settled decisively by Donald. After earlier partial results [**Ep65**, Corollary 1], [**KK80**, Proposition 6.1] and [**GL83**, Theorem 3.4], Donald showed that the only connected sums of lens spaces which embed are sums of those embeddable via the construction of Theorem 3.6.

THEOREM. [**Do15**, Theorem 1.1] *Let $L = \#_{i=1}^n L(p_i, q_i)$ be a connected sum of lens spaces. Then L embeds smoothly in S^4 if and only if each p_i is odd and there is a closed 3-manifold N such that $L \cong N \# -N$.* □

We mention two other aspects of embeddings of 3-manifolds in S^4. There are 149 manifolds in the census of 3-manifolds built from at most 11 tetrahedra and with hyperbolic linking form. Of these, 41 are known to embed smoothly in S^4, and four more embed smoothly in some homotopy 4-sphere. The Rokhlin μ-invariant and Oszváth-Szábo d-invariant have been used to show that 67 of the remaining 104 cannot embed smoothly in S^4 [**BB22**]. (The d-invariant is a $\mathbb{Q}$-valued function d on the set of $Spin^c$ structures, defined by means of Heegaard-Floer Theory [**OS03**]. All $\mathbb{Q}$-homology 3-spheres have $Spin^c$ structures.)

Motivated by the prospect of detecting a counter-example to the smooth 4-dimensional Poincaré conjecture through its smooth hypersurfaces, Freedman found criteria for a closed 3-manifold M to embed smoothly in S^4 in terms of a duality property of Heegaard splittings of M. (The key property of the standard 4-sphere invoked here is that $S^4 \setminus \{*\}$ is smoothly a product $\mathbb{R}^3 \times \mathbb{R}$.) This duality property is at present difficult to apply, but does imply the hyperbolicity of ℓ_M [**Fr24**].

The earliest work on embeddings of 3-manifolds in other 4-manifolds may be [**Ka88**]. Kawauchi considers embeddings of 3-manifolds M with $H_1(M)$ infinite and obtains bounds for signatures associated to infinite cyclic covers of M, in terms of the Betti numbers of a 4-manifold W with $M = \partial W$. Edmonds and

Livingston have studied the embeddings of lens spaces in $\#^n\overline{\mathbb{CP}}^2$. They show first that every lens space embeds smoothly in $\#^n\overline{\mathbb{CP}}^2$, for n sufficiently large, and then tabulate which lens spaces embed (smoothly or topologically) for n small, in some detail [**EL96**]. Some do not embed as locally flat submanifolds of $\#^4\overline{\mathbb{CP}}^2$, but all do so in $\#^8\overline{\mathbb{CP}}^2$ [**Ed05**]. For each $n > 0$ there are lens spaces that do not embed smoothly as a separating hypersurface into any smooth negative-definite 4-manifold W with $\beta_2(W) = n$ [**AMP22**]. This again uses the fact that lens spaces bound definite plumbings and the Diagonalization Theorem. However, if the lens space L is amphicheiral then it embeds in a 4-manifold W with $\beta_1(W) = 0$ and $\beta_1(W) = 2$ [**AMP24**]. Beyond this, little consideration has been given to 3-manifolds other than lens spaces in other 4-manifolds, according to [**AMP24**].

CHAPTER 6

3-Manifolds with Restrained Fundamental Group

In this chapter we shall show first that there are just thirteen 3-manifolds with restrained fundamental group which embed in homology 4-spheres, and all embed in S^4. The most difficult part of the argument involves consideration of 3-manifolds which are the union of two copies of the mapping cylinder N of the orientation cover of the Klein bottle. (This is an extension of the work in Chapter 4.) We then show that if there is a degree-1 map $f : M \to P$ which induces isomorphisms on $\mathbb{Z}[\pi_1 P]$-homology and $\pi_1 P$ is torsion-free and solvable, then M embeds in S^4 if and only if P does. (This result involves 4-dimensional surgery over $\pi_1 P$.)

6.1. Statement of the Main Result on Virtually Solvable Groups

If the fundamental group of a 3-manifold M is restrained then it is either finite or solvable, and M has one of the geometries $\mathbb{S}^3$, $\mathbb{S}^2 \times \mathbb{E}^1$, $\mathbb{E}^3$, $\mathbb{N}il^3$ or $\mathbb{S}ol^3$. In the first four cases M is a Seifert manifold.

The only infinite solvable 3-manifold group with torsion is D_∞, which we may exclude by the next result.

LEMMA 6.1. *If $\pi_1 M \cong D_\infty$ then M does not embed in any homology 4-sphere.*

PROOF. If $\pi_1 M \cong D_\infty$ then $M \cong \mathbb{RP}^3 \# \mathbb{RP}^3$, and so $\tau_M = H_1(M)$ is generated by elements x_1 and x_2 such that $\ell_M(x_i, x_i) \neq 0$. The lemma now follows immediately from Corollary 2.6.1. □

If M is flat, a $\mathbb{N}il^3$-manifold or $\mathbb{S}ol^3$-manifold and π/π' is infinite then M is the total space of a torus bundle over S^1, and conversely every such bundle has one of these geometries.

If M has one of these geometries and π/π' is finite then either M is the Hantzsche-Wendt flat manifold, or M is a $\mathbb{N}il^3$-manifold and is Seifert fibred over one of $S(2,2,2,2)$, $S(2,3,6)$, $S(2,4,4)$, $S(3,3,3)$ or $P(2,2)$, or M is a $\mathbb{S}ol^3$-manifold. The Hantzsche-Wendt flat manifold has fundamental group G_6 with presentation

$$\langle x, y \mid xy^2x^{-1} = y^{-2},\ yx^2y^{-1} = x^{-2}\rangle,$$

and $G_6/\langle\langle x^2, y^2\rangle\rangle \cong D_\infty$. The orbifold fundamental groups of $S(2,2,2,2)$ and $P(2,2)$ each map onto D_∞. If π is the group of a $\mathbb{S}ol^3$-manifold then $\sqrt{\pi} \cong \mathbb{Z}^2$, and so $\pi/\sqrt{\pi}$ is virtually $\mathbb{Z}$. Moreover, $\pi/\sqrt{\pi}$ has no non-trivial finite normal subgroup, and so $\pi/\sqrt{\pi} \cong D_\infty$, if π/π' is finite. In all cases, excepting only $\mathbb{N}il^3$-manifolds which have base orbifold $S(2,3,6)$, $S(2,4,4)$ or $S(3,3,3)$, if π/π' is finite then π maps onto D_∞, with kernel $\mathbb{Z}^2$. We then have $\pi \cong A *_C B$, where A and B are torsion-free, $[A : C] = [B : C] = 2$ and $C \cong \mathbb{Z}^2$. Since π/π' is finite the subgroups A and B cannot be $\mathbb{Z}^2$.

We may assume that A and B have presentations $\langle u, v \mid vuv^{-1} = u^{-1}\rangle$ and $\langle x, y \mid yxy^{-1} = x^{-1}\rangle$, respectively. Let $C < A$ have basis $\{u, v^2\}$. Then we may define a monomorphism $\phi : C \to B$ by $\phi(u) = x^a y^{2b}$ and $\phi(v^2) = x^c y^{2d}$, where $ad - bc = 1$. Let $G = A *_\phi B$ be the resulting generalized free product with amalgamation. Then G/G' is infinite if and only if $c = 0$. Otherwise, G is virtually abelian if $a = d = 0$, is virtually nilpotent if and only if exactly one of a, b or d is 0, and is solvable but not virtually nilpotent if all the entries are non-zero.

Such 3-manifolds have a corresponding decomposition as unions of two copies of the mapping cylinder N of the orientation cover of Kb. This 3-manifold N is orientable and $\partial N \cong T$, and $\pi_1 N = \pi_1 Kb$ has generators x and y such that $xyx^{-1c} = y^{-1}$. Fix a homeomorphism $h : \partial N \to T$ which carries x^2 and y to the standard basis of $\pi_1 T = \mathbb{Z}^2$. Then isotopy classes of self-homeomorphisms of ∂N correspond to automorphisms of $\mathbb{Z}^2$.

Every 3-manifold M with a solvable Lie geometry ($\mathbb{E}^3$, $\mathbb{N}il^3$ or $\mathbb{S}ol^3$) and such that $\pi_1 M$ maps onto D_∞ is a union $N \cup_\phi N$, for some self homeomorphism ϕ of N. It is easily seen that the automorphism $\left(\begin{smallmatrix} 1 & 0 \\ 0 & -1 \end{smallmatrix}\right)$ is induced by a self-homeomorphism of N, and so we may assume that $\det \phi = 1$.

We shall state our main result now, but defer the more difficult parts of the proof.

THEOREM 6.2. *If π is finite or solvable and M embeds in a homology 4-sphere then M is either the Poincaré homology sphere or is one of the twelve 3-manifolds listed above. More precisely, either*

(1) $\pi = 1$, $Q(8)$ *or* $I^* = SL(2,5)$ *and M is an $\mathbb{S}^3$-manifold; or*
(2) $\pi \cong \mathbb{Z}$ *and* $M \cong S^2 \times S^1$*; or*
(3) $\pi \cong \mathbb{Z}^3$ *or* $G_2 = \mathbb{Z}^2 \rtimes_{-I} \mathbb{Z}$*, and M is a flat 3-manifold; or*
(4) $\pi \cong \mathbb{Z}^2 \rtimes_A \mathbb{Z}$*, where* $A = \left(\begin{smallmatrix} 1 & 1 \\ 0 & 1 \end{smallmatrix}\right)$ *or* $A = \left(\begin{smallmatrix} -1 & 4 \\ 0 & -1 \end{smallmatrix}\right)$ *and M is a $\mathbb{N}il^3$-manifold;*
(5) $M \cong N \cup_\phi N$*, where* $\phi = \left(\begin{smallmatrix} 2 & -1 \\ 1 & 0 \end{smallmatrix}\right)$*, and M is a $\mathbb{N}il^3$-manifold; or*
(6) $M \cong N \cup_\phi N$*, where* $\phi = \left(\begin{smallmatrix} 2 & -3 \\ 1 & 2 \end{smallmatrix}\right)$, $\left(\begin{smallmatrix} 2 & -5 \\ 1 & -2 \end{smallmatrix}\right)$ *or* $\left(\begin{smallmatrix} 2 & -9 \\ 1 & -4 \end{smallmatrix}\right)$*, and M is a $\mathbb{S}ol^3$-manifold.*

PROOF. Suppose first that π is finite. Then it is a finite subgroup of $SO(4)$ which acts freely on S^3. If M embeds in S^4 then π/π' is a direct double, and so $\pi = 1$, I^* or is a generalized quaternionic group $Q(8k)$, where $k \geqslant 1$. The 3-sphere S^3 is the equator of S^4, while the homology sphere S^3/I^* embeds in S^4 by Freedman's result.

The quotients $S^3/Q(8k)$ with $k \geqslant 1$ are the total spaces of S^1 bundles over $\mathbb{RP}^2$ with normal Euler number $2k$. In particular, $S^3/Q(8)$ is the boundary of a regular neighbourhood of a smooth embedding of $\mathbb{RP}^2$ in S^4. However, none of the others embed. See Corollary 4.4.2.

If π has two ends then $\pi \cong \mathbb{Z}$, since D_∞ is excluded by Lemma 6.1. Hence $M \cong S^2 \times S^1$, which is the boundary of a regular neighbourhood of a smooth embedding of S^2 in S^4.

If π is not $\mathbb{Z}$ and π/π' is infinite then $\beta = \beta_1(M) = 1, 2$ or 3. If $\beta = 1$ then ρ is uniquely a semidirect product $\mathbb{Z}^2 \rtimes_A \mathbb{Z}$, for some $A \in SL(2,\mathbb{Z})$. Let $\Delta_A(t) = \det(A - tI)$ be the characteristic polynomial. Then $\Delta_A(1) \neq 0$, since $\beta = 1$. If $\Delta_A(-1) \neq 0$ also, then Kawauchi's Theorem applies, and $\Delta_A(t)$ is irreducible, giving a contradiction. Hence we must have $\Delta_A(t) = (t+1)^2$, and so $A = \left(\begin{smallmatrix} -1 & b \\ 0 & -1 \end{smallmatrix}\right)$ for some b. The mapping torus $T \rtimes_A S^1$ is also the total space of an S^1-bundle over Kb with Euler number b. This embeds in S^4 only if $b = 0$ or ± 4, by Corollary 4.4.2.

When $b = 0$ we have $\pi = \mathbb{Z}^2 \rtimes_{-I} \mathbb{Z}$ and M is the flat 3-manifold with holonomy of order 2. (The corresponding link is 8^3_9.) This is the boundary of a regular neighbourhood of a smooth embedding of Kb with normal Euler number 0. When $b = 4$ the corresponding link is 9^3_{19}.

If $\beta = 2$ then M is a $\mathbb{N}il^3$-manifold, and $\pi/\pi' \cong \mathbb{Z}^2 \oplus (\mathbb{Z}/q\mathbb{Z})$, for some $q \geqslant 1$. Clearly the torsion is a direct double only when $q = 1$. In this case M is the Heisenberg 3-manifold, with corresponding link 5^2_1.

If $\beta = 3$ then $\pi \cong \mathbb{Z}^3$ and M is the 3-torus, which embeds smoothly as the boundary of a regular neighbourhood of a smooth embedding of T in S^4 with a product neighbourhood. (The corresponding link is 6^3_1.)

We may assume henceforth that π has one end and π/π' is finite. The only such flat 3-manifold group is the group of the Hantzsche-Wendt flat 3-manifold with holonomy $(\mathbb{Z}/2\mathbb{Z})^2$. This does not embed as its torsion linking form is not hyperbolic, by Corollary 4.2.1.

If M is a $\mathbb{N}il^3$-manifold with Seifert base $S(3,3,3)$ and π/π' is a direct double then $M \cong M(0;(3,1),(3,1),(3,-1))$, with corresponding link 6^2_1.

The remaining cases all involve 3-manifolds of the form $M \cong N \cup_\phi N$. We defer discussion of these cases. □

Eleven of the twelve such 3-manifolds which embed in homology 4-spheres have presentations in terms of 0-framed bipartedly trivial links (and thus embed smoothly in S^4), namely six relatively familar links:

$U = 0_1$ (the unknot), giving $S^2 \times S^1$;
$Ho = 2^2_1$ (the Hopf link), giving S^3;
4^2_1 (the $(2,4)$-torus link), giving $S^3/Q(8)$;
$Wh = 5^2_1$ (the Whitehead link giving the $\mathbb{N}il^3$-manifold $M(1;(1,-1))$);
6^2_1 (the $(2,6)$-torus link), giving the $\mathbb{N}il^3$-manifold $M(0;(3,1),(3,1),(3,-1))$;
$Bo = 6^3_2$ (the Borromean rings), giving the 3-torus $S^1 \times S^1 \times S^1$;

and five which have no popular names: 8^3_9, 8^2_2, 9^3_{19}, 9^2_{53} and 9^2_{61}, giving the half-turn flat 3-manifold, two $\mathbb{N}il^3$-manifolds and two $\mathbb{S}ol^3$-manifolds, respectively. One further $\mathbb{S}ol^3$-manifold is given by a bipartedly ribbon link $(8_{20}, U)$.

The thirteenth 3-manifold with π restrained and which embeds is the Poincaré homology sphere S^3/I^*. However S^3/I^* does not embed smoothly (see Example 5.1), and so has no bipartedly slice link presentation.

Seven of the first twelve also arise as the boundaries of regular neighbourhoods of smooth embeddings of the 2-sphere S^2, the torus T, the projective plane $\mathbb{RP}^2$ or the Klein bottle Kb in S^4. Closed orientable surfaces embedded in S^4 have product neighbourhoods (normal Euler number 0), while the normal Euler number of an embedding of $\#^c\mathbb{RP}^2$ may have any value congruent to $2c$ *mod* (4) and between $-2c$ and $2c$ [**Ma69**]. (See also Corollary 4.4.2.)

6.2. Unions of Mapping Cylinders

The hard work is in dealing with the final cases. We need some further preparatory discussion and several lemmas before we can complete the proof of the main result (in Theorem 6.6).

THEOREM 6.3. *Let $M = N \cup_\phi N$, where N and $\phi = \left(\begin{smallmatrix} a & b \\ c & d\end{smallmatrix}\right)$ are as above. Then*

(1) *if $c = 0$ then ℓ_M is hyperbolic if and only if 4 divides b;*
(2) *if $c \neq 0$ then ℓ_M is hyperbolic if and only if b is odd, $c = 1$ and a and d are both even, but NOT both divisible by 4.*

PROOF. We shall identify the Klein bottle Kb with its image in N. Then the inclusion of $\partial N \cong T$ into $N \simeq Kb$ is homotopic to the orientation cover $\psi : T \to Kb$. The Klein bottle is also a twisted S^1 bundle over the circle. Let f and ζ be 1-cycles in Kb corresponding to a fibre and section of this bundle. We take $\xi = \psi^{-1}(\zeta)$ and a component f' of $\psi^{-1}(f)$ as 1-cycles in ∂N. Note that the bundle structure on Kb induces a bundle structure on N with typical fibres f and f' and section the mapping cylinder of $\pi|\xi$. This section is a Möbius band with centreline ζ and boundary ξ, which may be regarded as a 2-chain C with $\partial C = 2\zeta - \xi$. The Klein bottle itself may also be considered as a 2-chain with $\partial Kb = 2f$. Also, the fibres f and f' cobound an annular 2-chain A such that, putting $S = Kb + 2A$, we have $\partial S = 2f'$. Perturbing ζ and f' to intersect these 2-chains transversely, we see that $\zeta \bullet S = \pm 1$, $\zeta \bullet C = \pm 1$, $f' \bullet S = 0$ and $f' \bullet C = \pm 1$.

Now $H_1(N) \cong \langle x, y \mid 2y = 0\rangle$, where $x = [\zeta]$ and $y = [f']$, and $([\xi], [f']) = (2x, y)$ provides the standard framing of ∂N. Thus $H_1(M)$ has the presentation

$$\langle x_1, y_1, x_2, y_2 \mid 2y_1 = 2y_2 = 0,\ 2ax_1 + by_1 = 2x_2,\ 2cx_1 + dy_1 = y_2\rangle,$$

where we use subscripts to distinguish objects related to the two copies of N forming M. It is easily seen that if $c = 0$ then $H_1(M)$ is the direct sum of $\mathbb{Z}$ with a group of order 4, and the torsion subgroup is a direct double if and only if b is even, while if $c \neq 0$ then $H_1(M)$ is finite, of order $16c$, and is a direct double if and only if $c = 1$ and b is odd. We shall now determine ℓ_M in these two cases.

If $c = 0$ and b is even then $a = d = \pm 1$, and τ_M is generated by y_1 and $z = ax_1 - x_2$. The latter element is represented by the 1-cycle $\zeta' = a\zeta_1 - \zeta_2$. Under the identification map ϕ, the 1-cycles $a\xi_! + bf + 1'$ and ξ_2 are homologous and so bound a 2-chain $\Delta \subset \partial N$, and the chain $E = aC_1 - C_2 - \frac{1}{2}bS_1 + \Delta$ has boundary $2\zeta'$. Hence $\ell_M(y_1, y_1) = \frac{1}{2}(f_1' \bullet S_1) = 0$, $\ell_M(y_1, z) = \frac{1}{2}(f_1' \bullet E) = \frac{1}{2}(\zeta' \bullet S_1) = \frac{1}{2}$ and $\ell_M(z, z) = \frac{1}{2}(\zeta' \bullet E) = \pm\frac{1}{4}b$ in $\mathbb{Q}/\mathbb{Z}$, and (i) follows easily.

If $c = 1$ and b is odd then $H_1(M)$ is generated by x_1 and x_2. There are 2-chains $\Delta_1, \Delta_2 \subset \partial N$ such that the chains $G_1 = 2C_1 - dS_1 + S_2 + \Delta_1$ and $G_2 = 2C_2 - S_1 + aS_2 + \Delta_2$ have boundaries $4x_1$ and $4x_2$, respectively. (Here we have used both identifications $a\xi_1 + bf_1' \sim \xi_2$ and $\xi_1 + df_1' \sim f_2'$ due to ϕ, as well as the fact that $\det(\phi) = ad - bc = 1$.) Hence $\ell_M(x_1, x_1) = \frac{1}{4}(\zeta_1 \bullet G_1) = \frac{1}{4}(2 \pm d)$, $\ell_M(x_1, x_2) = \frac{1}{4}(\zeta_1 \bullet G_2) = \pm\frac{1}{4}$ and $\ell_M(x_2, x_2) = \frac{1}{4}(\zeta_2 \bullet G_2) = \frac{1}{4}(2 \pm a)$ in $\mathbb{Q}/\mathbb{Z}$, and (ii) follows easily. □

It follows from Theorem 6.3 that if $M = N \cup_\phi N$ embeds in a homology 4-sphere then either $c = 0$, in which case it is a torus bundle over S^1 and thus one of the manifolds listed in Theorem 6.2, or $c = 1$. In the second case we introduce the notation $M_{m,n}$ for the union corresponding to $\phi = \left(\begin{smallmatrix} m & mn-1 \\ 1 & n \end{smallmatrix}\right)$.

The group $\pi_1 M_{0,n}$ has the presentation $\langle x, u \mid x^2 = (xu^2)^2 = 1 \rangle$. The quotient by the infinite cyclic normal subgroup generated by x^2 is $\pi^{orb} P(2,2)$. Hence $M_{0,n}$ is Seifert fibred over $P(2,2)$, with generalized Euler invariant n. Thus $M_{0,n}$ has hyperbolic linking form if and only if $n \equiv 2 \; mod \; (4)$, by Corollary 4.2.1.

Let $N(e)$ denote the orientable S^1-bundle over $\mathbb{RP}^2$ with Euler number e, and let $N^*(e)$ denote the framed exterior of a fibre f in $N(e)$, with framing $([\mu], [f'])$ given by a meridian μ and a neighbouring fibre f'. For example, the mapping cylinder with the S^1-bundle structure and framing $([\xi], [f'] {=} (2x, y)$ used in the preceding proof is precisely the bundle $N^*(0)$. An alternative description for $M_{m,n}$ is $N \cup_{\phi'} N$, where $\phi' = \left(\begin{smallmatrix} m & 1-mn \\ 1 & -n \end{smallmatrix}\right)$. Now $\phi' = \left(\begin{smallmatrix} 1 & m \\ 0 & 1 \end{smallmatrix}\right)\left(\begin{smallmatrix} 0 & 1 \\ 1 & 0 \end{smallmatrix}\right)\left(\begin{smallmatrix} 1 & -n \\ 0 & 1 \end{smallmatrix}\right)$. Hence we may write

$$M_{m,n} = N^*(m) \cup_{\left(\begin{smallmatrix} 0 & 1 \\ 1 & 0 \end{smallmatrix}\right)} N^*(n).$$

That is, $M_{m,n}$ may be described by the plumbing diagram

as in [**Neu81**]. It is clear from this description that, firstly, $M_{m,n} \cong M_{n,m}$ and, secondly, reversing the orientation of $M_{m,n}$ corresponds to changing the sign of both m and n.

By using an argument similar to that for the proof of Theorem 4.4, we are able to put strong homological constraints on embeddings of the manifolds $M_{m,n}$, and hence complete the classification of those unions $N \cup_\phi N$ which embed.

Suppose that $M = M_{m,n}$ embeds in a homology 4-sphere Σ, with complementary regions X and Y. Then each summand of the finite group $H_1(M) \cong H_1(X) \oplus H_1(Y)$, is self-annihilating with respect to ℓ_M, by Lemma 2.6. We shall relabel the generators x_1, y_1, x_2, y_2 as x, y, u, v, respectively. Then $H_1(M) \cong (\mathbb{Z}/4\mathbb{Z})^2$, with generators x and u. We also have $y = 2u$ and $v = 2x$. By the proof of Theorem 6.3 we may write the matrix for ℓ_M with respect to the basis $\{x, u\}$ as

$$\ell_M = \begin{pmatrix} \frac{1}{4}(2+m) & \frac{1}{4} \\ \frac{1}{4} & \frac{1}{4}(2+n) \end{pmatrix} = \begin{pmatrix} 0 & \frac{1}{4} \\ \frac{1}{4} & \frac{1}{4}(2+n) \end{pmatrix}, \quad \text{if } m \equiv 2 \; mod \; (4).$$

Thus, in the case that $m \equiv n \equiv 2 \; mod \; (4)$, there are four possible self-orthogonal decompositions for $H_1(X) \oplus H_1(Y)$:

$$\langle x \rangle \oplus \langle u \rangle, \;\; \langle x \rangle \oplus \langle u + 2x \rangle, \;\; \langle x + 2u \rangle \oplus \langle u \rangle, \;\text{or}\; \langle x + 2u \rangle \oplus \langle u + 2x \rangle.$$

If $m \equiv 2 \ mod \ (4)$ and $n \equiv 0 \ mod \ (4)$ the four possibilities are

$$\langle x\rangle \oplus \langle u-x\rangle,\ \langle x\rangle \oplus \langle u+x\rangle,\ \langle -x+2u\rangle \oplus \langle u-x\rangle,\ \text{or}\ \langle -x+2u\rangle \oplus \langle u+x\rangle.$$

LEMMA 6.4. *If $M = M_{m,n}$ embeds in a homology 4-sphere Σ and $m \equiv 2 \ mod \ (4)$ then $m = \pm 2$.*

PROOF. The maps $H_1(Y) \cong \mathbb{Z}/4\mathbb{Z} \to \mathbb{Z}/2\mathbb{Z}$ determine 2-fold coverings $\widetilde{Y}' \to \widetilde{Y} \to Y$ which induce coverings $\widetilde{M}' \to \widetilde{M} \to M$ on the boundary corresponding to the maps

$$H_1(M) \xrightarrow{p_Y} \mathbb{Z}/4\mathbb{Z} \to \mathbb{Z}/2\mathbb{Z},$$

where p_Y is the projection onto $H_1(Y)$. Now given any one of the eight possible self-orthogonal decompositions listed above, these maps are such that

$$x \mapsto 0 \text{ or } 2 \mapsto 0, \quad y = 2u \mapsto 2 \mapsto 0,$$

$$u \mapsto \pm 1 \mapsto 1, \qquad v = 2x \mapsto 0 \mapsto 0.$$

Now $M = N^*(m) \cup N^*(n)$, and we may write $\widetilde{M} = \widetilde{N}^*(m) \cup \widetilde{N}^*(n)$. The piece $\widetilde{N}^*(m)$ is determined by the map

$$H_1(N^*(m)) \xrightarrow{i_*} H_1(M) \to \mathbb{Z}/2\mathbb{Z}.$$

But since $H_1(N^*(m))$ is generated by x and y, this map is trivial. Consequently $\widetilde{N}^*(m)$ consists of two disjoint copies of $N^*(m)$. On the other hand, we can see that since u (represented by an orientation reversing path in the base space of $N^*(n)$) is mapped to 1, while v (represented by a fibre of $N^*(n)$) is mapped to 0, then $\widetilde{N}^*(n)$ is just a framed S^1-bundle over the annulus, or $M(0;(1,2n))$ with two regular fibres removed. Using the plumbing notation of [**Neu81**], we have

$\widetilde{M} =$ 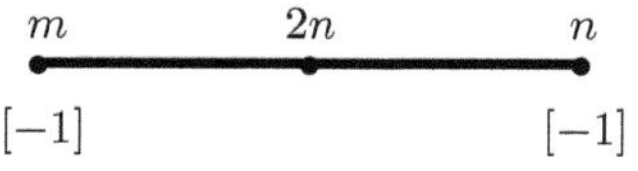

.

The boundary of $N^*(m)$ has framing $([\mu],[f])$ where $[\mu] = my + 2x$ and $[f] = y$. Since $y \mapsto 2$ under the map onto $\mathbb{Z}/4\mathbb{Z}$, in the covering map $\widetilde{M}' \to \widetilde{M}$ each copy of $N(m)$ is covered by $N(\frac{1}{2}m)$, and the covering projection θ acts by a half-turn rotation of each fibre. On the other hand, μ is identified (in $\widetilde{M}$) with a fibre of $\widetilde{N}^*(n)$, and since $[\mu] \mapsto 0$ in $\mathbb{Z}/4\mathbb{Z}$ (since m is even), this piece is itself covered by another S^1-bundle over the annulus, S_A, where θ acts on the base. Thus we have the plumbing description

$\widetilde{M}' =$ 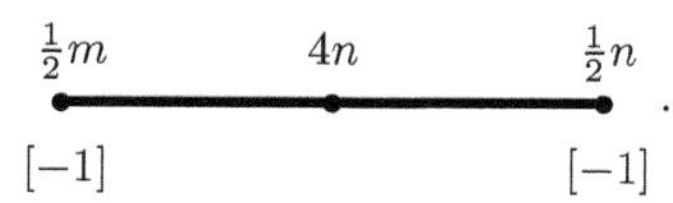

.

Note that since θ acts on the base of S_A, it extends to a fixed-point-free involution of the associated disc bundle D_A. Thus writing $W^0 = \widetilde{Y}' \cup_{S_A} D_A$, we see that the covering involution of $\widetilde{Y}'$ extends to a fixed-point-free involution θ^0 on W^0. Also observe that $\partial W^0 = N(\frac{m}{2}) \sqcup N(\frac{m}{2})$ (disjointly), since attaching the discs of

D_A corresponds to a trivial filling of the boundary of each $N^*(\frac{m}{2})$ piece. Writing $DN(\frac{m}{2})$ for the disc bundle associated to $N(\frac{m}{2})$, we may now construct the closed 4-manifold

$$W = W^0 \cup_{N(\frac{m}{2}) \sqcup N(\frac{m}{2})} (DN(\frac{m}{2}) \sqcup DN(\frac{m}{2})).$$

Now θ^0 extends to an involution θ_W of W with fixed point set the disjoint union of two copies of $\mathbb{RP}^2$, and the total normal Euler number of the fixed point set is $\frac{m}{2} + \frac{m}{2} = m$. Thus by the $\mathbb{Z}/2\mathbb{Z}$-Index Theorem, $\mathrm{Sign}(\theta_W, W) = m$. We now calculate $\beta_2(W)$.

Since $H_1(M)$ is finite it is a $\mathbb{Q}$-homology sphere, and so the complementary regions$X$ and Y are $\mathbb{Q}$-homology balls. In particular, $\chi(Y) = 1$. Hence $\chi(W^0) = \chi(\widetilde{Y}') = 4\chi(Y) = 4$, since $\chi(S_A) = \chi(T) = 0$ and $\chi(D_A) = \chi(S^1) = 0$. Thus $\chi(W) = \chi(W^0) + 2\chi(DN(\frac{m}{2})) = 4 + 2\chi(\mathbb{RP}^2) = 6$. But by Poincar'e duality this gives $\beta_2(W) = 4 + 2\beta_1(W)$.

By Lemma 4.7 the fact that $H_1(Y) = \mathbb{Z}/4\mathbb{Z}$ implies that $\beta_1(\widetilde{Y}') = 0$. Also, since the map on H_1 induced by the inclusion of an S^1-bundle into its associated disc bundle is an epimorphism, a Mayer-Vietoris argument shows that attaching disc bundles, as we have, does not increase β_1. Thus $\beta_1(W) = 0$ as well, and so $\beta_2(W) = 4$. Since $|\mathrm{Sign}(\theta_W, W)|$ is bounded above by this rank, we must have $m = \pm 2$. □

LEMMA 6.5. *If $M = M_{m,n}$ embeds in a homology 4-sphere Σ and $n \equiv 0$ mod (4) (and so $m \equiv 2$ mod (4) necessarily) then $m + n = \pm 2$.*

PROOF. This time we consider the 2-fold coverings $\overline{X}' \to \overline{X} \to X$ which induce coverings $\overline{M}' \to \overline{M} \to M$ corresponding to the maps

$$H_1(M) \xrightarrow{p_X} \mathbb{Z}/4\mathbb{Z} \to \mathbb{Z}/2\mathbb{Z}.$$

Here p_X is the projection onto the first factor in one of the four posssible self-orthogonal decompositions listed for the case $m \equiv 2$ *mod* (4) and $n \equiv 0$ *mod* (4). Thus the above maps are given by

$$x \mapsto 1 \mapsto 1, \quad y = 2u \mapsto 2 \mapsto 0,$$
$$u \mapsto \pm 1 \mapsto 1, \qquad v = 2x \mapsto 2 \mapsto 0.$$

Again $M = N^*(m) \cup N^*(n)$ and we write. $\overline{M} = \overline{N}^*(m) \cup \overline{N}^*(n)$. This time, since both x and u are mapped to 1 while $y, v \mapsto 0$ in $\mathbb{Z}/2\mathbb{Z}$, both $\overline{N}^*(m)$ and $\overline{N}^*(n)$ are S^1-bundles over annuli, and $\overline{M}$ has plumbing diagram

That is, $\overline{M} = N^{**}(2m) \cup_{T_a \sqcup T_b} N^{**}(2n)$, where the piece $N^{**}(2m)$ is just the framed exterior of a pair of fibres f_a and f_b in the bundle $M(0; (1, 2m))$ with Euler number m, and similarly for $N^{**}(2n)$. The boundary between the two pieces is a disjoint union $T_a \sqcup T_b$ of tori with framings (y_a, v_a) and (y_b, v_b), respectively, such that when $T_a \sqcup T_b$ covers the torus T in $M = N^*(m) \cup N^*(n)$ we have $y_a, y_b \mapsto y$ and

$v_a, v_b \mapsto v$. The framings of T_a and T_b given here correspond to the usual framings of f_a and f_b, where for $N^{**}(2m)$ the fibres represent y_a and $-y_b$, while for $N^{**}(2n)$ the fibres represent v_a and $-v_b$, respectively.

Now, since both $y, v \mapsto 2$ in $\mathbb{Z}/4\mathbb{Z}$, both the fibre of $N^{**}(2m)$ and the fibre of $N^{**}(2n)$ are "unwrapped" in the covering $\overline{M}' \to \overline{M}$. Thus, $\overline{M}' = N^{**}(m) \cup_{T_a \sqcup T_b} N^{**}(n)$, and we may now view $\overline{X}'$ as a 4-manifold with (covering) involution θ which acts on each fibred piece of the boundary by a rotation of the fibres through a half-turn. Observe also that the behaviour of θ on a neighbourhood $(T'_a \cup T'_b) \times [-1, 1]$ of $T'_a \sqcup T'_b$ is independent of m and n. Thus, if we begin with $M_1 = M_{m+1,n_1}$ in Σ_1 and $M_2 = M_{m_2,n_2}$ in Σ_2, each of which satisfies our hypotheses, and construct $\overline{X}'_1$ and $\overline{X}'_2$ as above, and define $W^0 = \overline{X}'_1 \cup_{(T'_a \cup T'_b)\times[-1,1]} -\overline{X}'_2$, then the covering involutions θ_1 and θ_2 behave identically on $(T'_a \cup T'_b) \times [-1, 1]$, and together define a fixed-point-free involution θ^0 of W^0. Also ∂W^0 is the disjoint union of S^1-bundles over tori $M(1; (1, m_1 - m_2)) \sqcup M(1; (1, (n_1 - n_2))$, and θ acts on each boundary component by a half-turn rotation of the fibre. Attaching the associated disc bundles, we obtain a closed 4-manifold W with involution θ which has fixed-point set $F = T_x \sqcup T_u$ with total normal Euler number $e(F) = (m_1 - m_2) + (n_1 - n_2)$. Now, supposing that $M = M_{m,n}$ embeds in Σ, we may take $M_1 = M$ and $M_2 = -M = M_{-m,-n}$ (embedded in $-\Sigma$) so that, by the $\mathbb{Z}/2\mathbb{Z}$-Index Theorem,

$$\mathrm{sign}(\theta, W) = e(F) = 2(m + n).$$

Consequently, $|m + n| \leqslant \frac{1}{2}\beta_2(W)$.

We have $\chi(W) = \chi(W^0) = 2\chi(\overline{X'}) = 8\chi(X) = 8$, by arguments similar to those of Lemma 6.4. Poincaré duality gives $\beta_2(W) = 6 + 2\beta_1(W)$. But $\beta_1(W) \leqslant \beta_1(W^0) + 1$, by a Mayer-Vietoris argument, since once again $\beta_1(\overline{X}') = 0$. Thus $\beta_2(W) = 8$, and so $|m + n| \leqslant 4$. Hence we must have $m + n = \pm 2$. □

We may now complete the proof of Theorem 6.2.

THEOREM 6.6. *If $M = N \cup_\phi N$ embeds in a homology 4-sphere then either it is a torus bundle or ϕ is one of $\left(\begin{smallmatrix} 2 & -1 \\ 1 & 0 \end{smallmatrix}\right)$, $\left(\begin{smallmatrix} 2 & -3 \\ 1 & 2 \end{smallmatrix}\right)$, $\left(\begin{smallmatrix} 2 & -5 \\ 1 & -2 \end{smallmatrix}\right)$ or $\left(\begin{smallmatrix} 2 & -9 \\ 1 & -4 \end{smallmatrix}\right)$.*

PROOF. It is clear from Theorem 6.3 that $\beta_1(M) > 0$ if and only if $c = 0$, and then M is a torus bundle. If $c \neq 0$ and ℓ_M is hyperbolic then $c = 1$ and $M = M_{m,n}$ for some even integers m, n, which are not both divisible by 4. Since $M_{n,m} \cong M_{m,n}$, we may assume that $m \equiv 2$ *mod* (4). We then have $m = \pm 2$ and either $n = \pm 2$ or $n \equiv 0$ *mod* (4). In the latter case $m + n = \pm 2$, by Lemma 6.5. After changing the orientation, if necessary, we may assume that $m = 2$, and so $n = 0$ or -4. □

6.3. A Link Presentation for $M_{2,4}$

In Figure 6.1 we have redrawn the diagram from [**CH98**, Figure 1] so that the non-trivial component is visibly a ribbon knot. In [**CH98**] the Kirby calculus was used to show that $M_{2,4} \cong M(L)$. Here we shall instead use the Wirtinger presentation for the link group πL to show that $\pi_1 M(L) \cong \pi_1 M_{2,4}$. (The group $\pi_1 M(L)$ is the quotient of πL by the normal closure of the longitudes.)

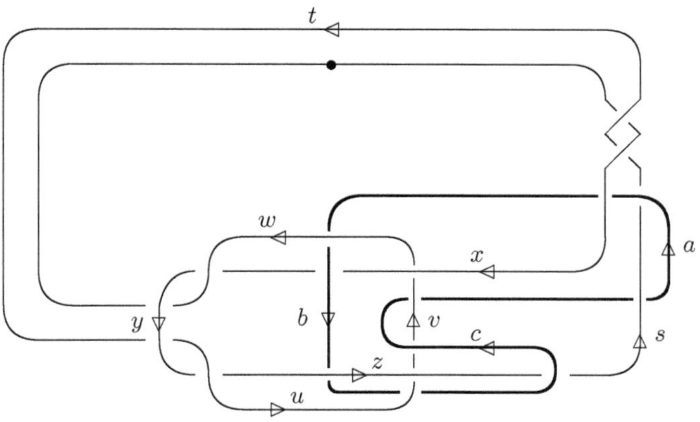

Figure 6.1 $L = (8_{20}, U)$.

The group πL has the Wirtinger presentation

$$\langle a, b, c, s, t, u, v, w, x, y, z \mid axtx^{-1}a^{-1} = s,\ bwyw^{-1}b^{-1} = x,$$
$$csc^{-1} = z,\ uyu^{-1} = z,\ vsas^{-1}v^{-1} = c,\ wx^{-1}axw^{-1} = b,\ xwx^{-1} = v,$$
$$ytxt^{-1}y^{-1} = w,\ yty^{-1} = u,\ zucu^{-1}z^{-1} = b,\ zuz^{-1} = cvc^{-1}.\rangle$$

The longitudes associated with the meridianal generators a and x are $\ell_a = xw^{-1}zuvs$ and $\ell_s = axy^{-1}z^{-1}cxytbu^{-1}cs^{-1}$. We may use the relations $z = csc^{-1} = uyu^{-1}$, $v = xwx^{-1}$ and $u = yty^{-1}$ to eliminate these generators, to obtain the presentation

$$\langle a, b, c, s, t, w, x, y, \mid axt = sax,\ bwy = xbw,\ ytx = wyt,\ ytc = byt,$$
$$csc^{-1} = ytyt^{-1}y^{-1},\ xwx^{-1}sas^{-1}xw^{-1}x^{-1} = c,\ wx^{-1}axw^{-1} = b,$$
$$sc^{-1}yty^{-1}cs^{-1} = xwx^{-1}\rangle.$$

Let $B = bw$ and $E = yt$. Then πL has the presentation

$$\langle a, B, c, s, w, x, y, E \mid axy^{-1}E = sax,\ By = xB,\ Ex = wE,$$
$$Ec = Bw^{-1}E,\ csc^{-1} = EyE^{-1},\ xwx^{-1}sas^{-1}xw^{-1}x^{-1} = c,$$
$$wx^{-1}ax = B,\ sc^{-1}Ey^{-1}cs^{-1} = xwx^{-1}\rangle.$$

The longitudes are now $\ell_a = xw^{-1}zuvs = xw^{-1}Exwx^{-1}s$ and $\ell_s = axy^{-1}z^{-1}cxytbwu^{-1}cs^{-2} = axy^{-1}cs^{-1}xEByE^{-1}cs^{-2}$.

Eliminating $y = B^{-1}xB$ and adjoining the relations $\ell_a = \ell_s = 1$ gives a presentation for $\pi = \pi_1 M(L)$.

$$\langle a, B, c, s, w, x, E \mid axB^{-1}x^{-1}BE = sax,\ Ex = wE,\ Ec = Bw^{-1}E,$$
$$csc^{-1} = EB^{-1}xBE^{-1},\ xwx^{-1}sas^{-1}xw^{-1}x^{-1} = c,\ a = xw^{-1}Bx^{-1},$$
$$sc^{-1}EB^{-1}x^{-1}Bcs^{-1} = xwx^{-1},\ xw^{-1}Exwx^{-1}s = 1,$$
$$axB^{-1}x^{-1}Bcs^{-1}xExBE^{-1}cs^{-2} = 1\rangle.$$

Eliminating $a = xw^{-1}Bx^{-1}$ gives

$$\langle B, c, s, w, x, E \mid xw^{-1}x^{-1}BE = sxw^{-1}B,\ Ex = wE,\ Ecx = BE,$$
$$csc^{-1} = EB^{-1}xBE^{-1},\ BE = cxwx^{-1}sx,\ Exwx^{-1}sx = w,$$
$$sc^{-1}EB^{-1}x^{-1}Bcs^{-1} = xwx^{-1},\ xw^{-1}x^{-1}Bcs^{-1}xExBE^{-1}cs^{-2} = 1\rangle.$$

The second and sixth relations give $Ewx^{-1}sx = 1$, and with the first we get $BE = xwx^{-1}sxw^{-1}B = xE^{-1}w^{-1}B = E^{-1}B$. Using the fourth and sixth relations, the seventh simplifies to $c^{-1}Ec = xwx^{-1}s = xE^{-1}x^{-1}$, or $EcxE = cx$, and so the presentation now becomes

$$\langle B, c, s, w, x, E \mid BE = E^{-1}B,\ Ex = wE,\ Ecx = BE,$$
$$csc^{-1} = EB^{-1}xBE^{-1},\ BE = cxwx^{-1}sx,\ Ewx^{-1}sx = 1,$$
$$EcxE = cx,\ xw^{-1}x^{-1}Bcs^{-1}xExBE^{-1}cs^{-2} = 1\rangle.$$

Eliminating $w = ExE^{-1}$ gives:

$$\langle B, c, s, x, E \mid BE = E^{-1}B,\ Ecx = BE,\ csc^{-1} = EB^{-1}xBE^{-1},$$
$$BE = cxExE^{-1}x^{-1}sx,\ xE^{-1}x^{-1}sx = E^{-2},$$
$$xEx^{-1}E^{-1}x^{-1}Bcs^{-1}xExBE^{-1}cs^{-2} = 1 :\ cx = BE^2,\ Ecx = cxE^{-1}\rangle,$$

in which the last two relations are redundant. Applying the fifth relation to the fourth gives $BE = cxE^{-1}$ or $BE^2 = cx$, and rewriting the sixth as

$$E^{-1}x^{-1}Bcs^{-1}xExBE^{-1}cs^{-1} = xE^{-1}x^{-1}s = E^{-2}x^{-1}$$

gives

$$\langle B, c, s, x, E \mid BE = E^{-1}B,\ Ecx = BE,\ s = c^{-1}EB^{-1}xBE^{-1}c,$$
$$xE^{-1}x^{-1}sx = E^{-2},\ Ex^{-1}Bcs^{-1}xExBE^{-1}cs^{-1}x = 1\rangle.$$

Eliminating $s = c^{-1}EB^{-1}xBE^{-1}c$ gives

$$\langle B, c, x, E \mid BE = E^{-1}B,\ Ecx = BE,$$
$$xE^{-1}x^{-1}c^{-1}EB^{-1}xBE^{-1}cx = E^{-2},\ Ex^{-1}E^{-1}x^{-1}BE^{-1}cxE^2Bcx = 1\rangle.$$

Eliminating $c = BE^2x^{-1}$ and using $BE = E^{-1}B$ to simplify gives

$$\langle B, x, E \mid BE = E^{-1}B,\ xB^{-2}E^{-4}xB^2E^5 = 1,\ x^{-1}E^{-1}x^{-1}B^4E^8 = 1\rangle.$$

Since $xB^2E^4x^{-1} = xB^2E^5x(xEx)^{-1} = B^{-2}E^{-4}$ and hence
$x^2E = xB^4E^8x^{-1} = B^{-4}E^{-8}$, this becomes

$$\langle B, x, E \mid BE = E^{-1}B,\ x^2 = B^{-4}E^{-9},\ xEx^{-1} = B^8E^{17},$$
$$xB^2E^4x^{-1} = (B^2E^4)^{-1}\rangle.$$

We shall now relabel the generators to bring out the similarities with the coordinates used in describing $N \cup_\phi N$. Let $X = E, Y = B^{-1}, U = B^{-2}E^{-4}$ and $V = x$. Then we have the presentation

$$\langle X, Y, U, V \mid XYX^{-1} = Y^{-1},\ UVU^{-1} = V^{-1},$$
$$U^2 = (X^2)^2Y^{-9},\ V = (X^2)Y^{-4}\rangle,$$

and so $\pi \cong \pi_1 M_{2,4} = \pi_1 N *_\phi \pi_1 N$, where $\phi = \left(\begin{smallmatrix} 2 & -9 \\ 1 & -4 \end{smallmatrix}\right)$.

6.4. Homologically Related 3-Manifolds

The question of which homology 3-spheres embed smoothly in S^4 is delicate, and of continuing interest. However, Freedman showed that the corresponding question for TOP locally flat embeddings has a dramatically simpler answer: every homology 3-sphere embeds in S^4 [**FQ**, Corollary 9.3c]. Inspired by this, we may consider the question of locally flat embeddings of 3-manifolds with a given homology type. A homology 3-sphere is a 3-manifold Σ with a map $f : \Sigma \to S^3$ which induces an isomorphism on integral homology. When the model 3-manifold is not S^3, we must use *local* coefficients in formulating the appropriate homological relation. For instance, the manifolds $M(K)$ obtained by 0-framed surgery on knots K in S^3 admit maps $f : M(K) \to S^2 \times S^1$ which induce isomorphisms on homology with simple coefficients $\mathbb{Z}$. If the knot K is not algebraically slice then $M(K)$ does not even admit a Poincaré embedding into S^4. On the other hand, if f induces isomorphisms on homology with local coefficients $\mathbb{Z}[\mathbb{Z}]$ then (as we shall see) $M(K)$ embeds in S^4. Our strategy involves 4-dimensional surgery, and so we must assume that the fundamental group of the model manifold is "good" in the sense of [**FQ**]. At the time of writing, all known good 3-manifold groups are either solvable or finite.

We recall the statement of Lemma 2.7: *If M and M' are $\mathbb{Z}$-homology cobordant 3-manifolds then M embeds in a homology 4-sphere if and only if M' embeds in a (possibly different) homology 4-sphere.*

By retaining more control over the fundamental group we shall be able to refine the argument for Lemma 2.7 to obtain embeddings of certain 3-manifolds in S^4. However, Lemma 2.7 may be used as it stands to obtain non-embedding results.

LEMMA 6.7. *Let Y be a connected cell complex such that $H_2(Y;\mathbb{Z}) \cong \mathbb{Z}$ and $H_i(Y;\mathbb{Z}) = 0$ for $i \neq 0$ or 2, and let $k > 1$. Suppose that $\mathbb{Z}/k\mathbb{Z}$ acts freely on Y, and that the induced action on $H_2(Y;\mathbb{Z})$ is trivial. Let X be the quotient space. Then $H_p(X;\mathbb{Z}) \neq 0$ for all odd $p > 0$.*

PROOF. The homology spectral sequence for the covering $Y \mapsto X$ has E^2 page

$$E^2_{p,q} = H_p(\mathbb{Z}/k\mathbb{Z}; H_q(Y;\mathbb{Z})) \Rightarrow H_{p+q}(X;\mathbb{Z})$$

with only two non-zero rows. All entries with p even are 0, except for $E^2_{0,0} = \mathbb{Z}$ and $E^2_{0,2} \cong \mathbb{Z}$, while all other non-zero entries are cyclic of order k. Hence all the differentials are 0, so $E^\infty_{p,q} = E^2_{p,q}$ for all p, q, and $H_p(X;\mathbb{Z}) \neq 0$ for all odd $p > 0$. □

Of course $\mathbb{Z}/2\mathbb{Z}$ acts freely on S^2 with quotient $\mathbb{RP}^2$, but the action on $H_2(S^2;\mathbb{Z})$ is non-trivial. We shall apply this lemma in the next, with Y and X infinite covering spaces of a 3-manifold.

LEMMA 6.8. *Let M be a 3-manifold and ν a perfect normal subgroup of $\pi = \pi_1 M$ such that $\rho = \pi/\nu$ has finitely many ends. Then either*

(1) *ρ is finite and has cohomological period dividing 4; or*
(2) *$\rho \cong \mathbb{Z}$ or D_∞; or*
(3) *ρ is a PD_3^+-group.*

PROOF. Let M_ν be the covering space associated to ν.

If ρ is finite then M_ν is a homology sphere, since ν is perfect. Since ρ acts freely on M_ν it has cohomological period dividing 4.

If ρ has two ends then it has a finite normal subgroup F such that $\rho/F \cong \mathbb{Z}$ or D_∞. Moreover, M_ν has the homology of S^2. Hence the subgroup of $\rho = Aut(M/M_\nu)$ that acts trivially on $H_2(M_\nu) \cong \mathbb{Z}$ has index at most 2. If g is an element of finite order in this subgroup then the quotient of M_ν by the action of g is an open 3-manifold, and so we may apply Lemma 6.7 to conclude that $g = 1$. Hence $\rho \cong \mathbb{Z}$, $\mathbb{Z} \oplus (\mathbb{Z}/2\mathbb{Z})$ or D_∞.

Suppose that $\rho \cong \mathbb{Z} \oplus (\mathbb{Z}/2\mathbb{Z})$. Let $g \in \pi$ represent the element of order 2 in ρ. Then g acts non-trivially on $H_2(M_\nu)$, by Lemma 6.7. Hence the action of ρ has kernel $\cong \mathbb{Z}$. Let M^+ be the 2-fold covering space of M with fundamental group $\nu \rtimes \mathbb{Z}$ the preimage of this kernel in π. Then $H_1(M^+) \cong \mathbb{Z}$, and so $H_i(M^+) \cong \mathbb{Z}$ for $i \leqslant 3$. Moreover, it is easily seen that the map $M_\nu \mapsto M^+$ induces an isomorphism on H_2. Since g acts trivially on $H_1(M^+)$ and non-trivially on $H_2(M^+)$ it must be orientation-reversing, contrary to our standing hypothesis that M be orientable. Therefore, we may exclude this possibility.

If ρ has one end then M_ν is acyclic. We may assume that M_ν has the ρ-equivariant triangulation lifted from a triangulation of M. The cellular chain complex of M_ν is now a finitely generated free $\mathbb{Z}[\rho]$-complex, of length 3. Since it is a resolution of the augmentation $\mathbb{Z}[\rho]$-module $\mathbb{Z}$, we see that $c.d.\rho \leqslant 3$. The universal coefficient spectral sequence for M with coefficients $\mathbb{Z}[\rho]$ collapses to give isomorphisms $H^i(\rho; \mathbb{Z}[\rho]) \cong H^i(M; \mathbb{Z}[\rho])$. As this is in turn isomorphic to $H_{3-i}(M; \mathbb{Z}[\rho]) = H_{3-i}(M_\nu)$, by Poincaré duality, ρ is a PD_3^+-group. □

This lemma is enough for our purposes, since virtually solvable groups have finitely many ends.

LEMMA 6.9. *Let M be a 3-manifold such that $\pi = \pi_1 M$ is an extension of a torsion-free solvable group ρ by a perfect normal subgroup ν. Then either $\rho = 1$ and M is a homology sphere, or $\rho \cong \mathbb{Z}$ and there is a Λ-homology isomomorphism from M to $S^2 \times S^1$, or ρ is the fundamental group of an aspherical 3-manifold P and there is a $\mathbb{Z}[\rho]$-homology isomorphism from M to P.*

PROOF. Since ρ is torsion-free and has finitely many ends it is either 1, $\mathbb{Z}$ or is a PD_3^+-group. The result is clear if $\rho = 1$.

If $\rho = \mathbb{Z}$ then there is a $\mathbb{Z}$-homology equivalence $h : M \to S^2 \times S^1$, by Lemma 1.1. Let $\tilde{h} : M_\nu \to S^2 \times \mathbb{R}$ be a lift of h to the infinite cyclic covering spaces. Since ν is perfect it follows easily from the Wang sequence for the covering projection that $\tilde{h}$ is a homology isomorphism, and hence that h is a $\mathbb{Z}[\rho]$-homology isomorphism.

Suppose now that ρ is a PD_3^+-group. As it is virtually solvable it is virtually polycyclic, of Hirsch length 3 [**Bie**, Theorem 9.23]. Hence it is the fundamental group of a 3-manifold P, which is either flat (if ρ is virtually abelian), a $\mathbb{N}il^3$-manifold (if ρ is virtually nilpotent but not virtually abelian) or a $\mathbb{S}ol^3$-manifold (if ρ is not virtually nilpotent). In all cases, every automorphism of ρ is realizable by some based self-homeomorphism of P, and so the epimorphism from π to ρ may be realized by a map $h : M \to P$. Let σ be a normal subgroup of finite index in ρ which is a poly-$\mathbb{Z}$ group, and let $\lambda : \sigma \to \mathbb{Z}$ be an epimorphism. Then $\kappa = \mathrm{Ker}(\lambda)$ is poly-$\mathbb{Z}$ of Hirsch length 2, and so is isomorphic to $\mathbb{Z}^2$, since ρ is orientable. Let $h_\kappa : M_\kappa \to P_\kappa \simeq S^1 \times S^1 \times \mathbb{R}$ be a lift of h to the covering spaces associated to κ and its preimage in π. The cohomology rings of M_κ and P_κ are generated in degree 1 and so h_κ induces a (co)homology isomorphism. It follows easily that h is a $\mathbb{Z}[\rho]$-homology isomorphism. □

In the final case P is either flat, a $\mathbb{N}il^3$-manifold or a $\mathbb{S}ol^3$-manifold.

6.5. Application of Surgery

The symbols M, P, ρ and h shall retain their meaning from Lemma 6.9 in the following lemmas, except that we shall allow P to denote S^3 or $S^2 \times S^1$ also, where appropriate.

LEMMA 6.10. *The maps h and id_P are normally cobordant.*

PROOF. The result is clear if $P = S^3$. Suppose that $P = S^2 \times S^1$. Then $\mathbb{Z}[\rho] \cong \Lambda$. There is a normal map $n : T = S^1 \times S^1 \to S^2$ with Arf invariant 1, and $n \times id_S^1$ is a normal map with non-trivial surgery obstruction in $L_3(\Lambda) \cong \mathbb{Z}/2\mathbb{Z}$. Thus, the surgery obstruction map $\sigma_3(S^2 \times S^1)$ from $\mathcal{N}(S^2 \times S^1) = \mathbb{Z}/2\mathbb{Z}$ to $L_3(\Lambda)$ is bijective.

If ρ is virtually polycyclic we may appeal to [**FJ88**] instead to see that $Wh(\rho) = 0$ and $\sigma_3(P) : \mathcal{N}(P) \to L_3(\mathbb{Z}[\rho])$ is an isomorphism.

Since in all cases h is a $\mathbb{Z}[\rho]$-homology isomorphism, it has trivial surgery obstruction and so is normally cobordant to id_P. □

LEMMA 6.11. *There is an H-cobordism W over $\mathbb{Z}[\rho]$ from M to P with $\pi_1 W = \rho$.*

PROOF. Suppose first that $\rho \cong \mathbb{Z}$. Let $F : N \to S^2 \times S^1 \times [0,1]$ be a normal cobordism from h to $id_{S^2 \times S^1}$. The surgery obstruction of F is determined by the signature of N, since the natural homomorphism from $L_4(\mathbb{Z})$ to $L_4(\mathbb{Z}[\mathbb{Z}])$ is an isomorphism. By forming the connected sum of F with an appropriate multiple of the degree-1 map from the E_8-manifold to S^4 we may obtain a normal cobordism with trivial surgery obstruction. If ρ is virtually polycyclic then we may appeal to [**FJ88**] instead, to see that $\sigma_4(P \times [0,1], \partial)$ is an isomorphism. Thus, we may in all cases assume that there is a normal cobordism from h to id_P with trivial surgery obstruction.

Since ρ is "good", we may apply [**FQ**, Theorem 11.3A] to complete surgery relative to the boundary to obtain a simple homotopy equivalence $f : W \to P \times [0,1]$, where $\partial W = M \sqcup P$, $f|_M = h$ and $f|_P = id_P$. Such a 4-manifold is clearly an H-cobordism over $\mathbb{Z}[\rho]$. □

The argument of the first paragraph of Lemma 6.10 together with Lemma 6.11 shows that any 3-manifold with the same homology as $S^2 \times S^1$ is normally cobordant to $S^2 \times S^1$. However, in general the normal cobordism cannot be improved to an H-cobordism over $\mathbb{Z}$. See [**FQ**, §11.8].

THEOREM 6.12. *Let P be a 3-manifold which embeds in S^4 and such that every automorphism of $\rho = \pi_1 P$ is induced by some (base-point-preserving) self-homeomorphism of P. Let W be a cobordism with $\partial W = P \sqcup M$ such that the inclusion of P into W induces an isomorphism $\rho \cong \pi_1 W$, and which is an H-cobordism over $\mathbb{Z}[\rho]$. Then M also embeds in S^4.*

PROOF. Let $Z = W \cup_M W$ be the union of two copies of W, doubled along M', and let j_1 and j_2 be the natural identifications of P with the two components of ∂Z. Since the inclusion of M into W induces an epimorphism on fundamental groups, the maps $\pi(j_i)$ are isomorphisms, for $i = 1, 2$, by Van Kampen's Theorem. Let ψ be a (basepoint-preserving) self-homeomorphism of P inducing the isomorphism $\pi(j_2)^{-1}\pi(j_1)$. Let X and Y be the complementary regions of an embedding of P in S^4, let $i_X : P \to \partial X$ and $i_Y : P \to \partial Y$ be the natural inclusions, and let $\Sigma = X \cup Z \cup Y$, where we identify $i_X(p)$ with $j_1(p)$ and $i_Y(p)$ with $j_2\psi(p)$, for all $p \in P$. Then Σ is simply connected and $\chi(\Sigma) = \chi(X) + \chi(Y) = \chi(S^4)$, so $\Sigma \cong S^4$. As Σ contains M as a locally flat submanifold this proves the theorem. □

There is a parallel argument on the homotopy level. If $h : M \to P$ is a $\mathbb{Z}[\rho]$-homology isomorphism where $\rho = \pi_1 P$, and $Z(h)$ is the mapping cylinder of h then $(Z(h); m, P)$ is a Poincaré duality triad with $\pi_1 Z(h) \cong \rho$. Hence if P has a Poincaré embedding in S^4 and automorphisms of ρ are realizable by self-homotopy equivalences of P then M also has a Poincaré embedding in S^4.

It remains possible that a model manifold P might not embed in S^4, while M does. However, we may settle the question for 3-manifolds with models having virtually solvable fundamental group.

COROLLARY 6.12.1. *If π is an extension of an infinite solvable group ρ by a perfect normal subgroup then M embeds in S^4 if and only if ρ is one of the torsion-free infinite groups corresponding to cases (2) to (6) of Theorem 6.2.*

PROOF. As observed in Lemma 6.9, solvable PD_3-groups are 3-manifold groups. The condition is necessary, by Theorem 6.2 and Lemma 2.7. In each case, every automorphism of ρ is induced by a self-homeomorphism of P, and P embeds smoothly in S^4. Therefore, the condition is also sufficient, by Lemmas 6.9, 6.10 and 6.11 and Theorem 6.12. □

The connected sum of a homology sphere with $S^2 \times S^1$ satisfies the hypotheses of Theorem 6.12, but in this case the embeddability follows immediately from the corresponding result for homology spheres. We may construct more interesting examples from knot manifolds. Let $M = M(K)$, where K is a knot with Alexander polynomial $\Delta_1(K) = 1$, and let $\pi = \pi_1 M(K)$. Then π' is perfect, and so M embeds, by the above corollary. If K is non-trivial then M is aspherical [**Ga87**], and thus is not a connected sum of $S^2 \times S^1$ with a homology sphere.

There is another argument for the case $\rho = \mathbb{Z}$. A 3-manifold M such that $\pi_1 M$ is an extension of $\mathbb{Z}$ by a perfect normal subgroup ν may be obtained by 0-framed surgery on a knot K in an homology 3-sphere Σ. Then $\Sigma = \partial C$ for some contractible 4-manifold C. Since ν is perfect K has Alexander polynomial 1, and so it bounds a disc D in C [**FQ**, Theorem 11.7B]. A mild variation on the 0-framed surgery construction shows that M embeds in $C \cup_{\Sigma} C$, which is a homotopy 4-sphere, and so is homeomorphic to S^4.

The elementary argument sketched above for the case $\rho \cong \mathbb{Z}$ may be extended to realize the other possibilities considered here by examples based on links obtained by tying Alexander polynomial 1 knots along various components of links representing the model 3-manifolds. We do not know whether there are other $\mathbb{Z}[\rho]$-homology equivalent 3-manifolds which cannot be handled in this way, but think it likely.

There is a similar result for the other torsion-free solvable groups corresponding to the manifolds listed in Theorem 6.2.

6.6. Extensions of the Argument

We may ask whether the present approach works for other groups ρ. When $\rho = D_\infty$ no such 3-manifold embeds, by Lemmas 2.7 and 6.1, and so the situation is clear in this case.

We are left with finite groups with cohomological period dividing 4. If such a group has abelianization a direct double then it is 1, a generalized quaternionic group $Q(8k)$ (with $k \geqslant 1$), the binary icosahedral group $I^* = SL(2,5)$, or one of the groups $Q(2^n a, b, c)$ (where a, b, c are odd and relatively prime, and either $n = 3$ and at most one of a, b, c is 1, or $n > 3$ and $bc > 1$). Groups of the latter class do not act freely on S^3 [**Orl**, Theorem 6.2], and so are not fundamental groups of 3-manifolds, by the Perel'man-Thurston Geometrization Theorem. Closed 3-manifolds with finite fundamental group are Seifert fibred, and the only finite groups realized by 3-manifolds which embed in S^4 are 1, $Q(8)$ and I^*, by Theorem 6.2. However, in so far as the argument for this result rests upon the $\mathbb{Z}/2\mathbb{Z}$-Index Theorem (and thus on the geometry of $\mathbb{Z}/2\mathbb{Z}$-actions on 4-manifolds) and is not purely homological, it remains possible that some 3-manifold M with fundamental group an extension of ρ by a perfect normal subgroup might embed, for ρ any one of these groups.

Since S/I^* is a homology 3-sphere it embeds in S^4. Although $L_4(I^*)$ has rank 9 there are no obstructions to embedding 3-manifolds with fundamental group an extension of I^* by a perfect normal subgroup, for this case may be subsumed into the case $\rho = 1$ settled by Freedman.

The spherical space form $S^3/Q(8)$ embeds smoothly in S^4 as the boundary of a regular neighbourhood of an embedding of $\mathbb{RP}^2$. Every automorphism of $Q(8)$ is realizable by a self-homeomorphism of $S^3/Q(8)$ [**Pr77**]. Let M be a closed 3-manifold such that $\pi = \pi_1 M$ is an extension of $Q(8)$ by a perfect normal subgroup ν, and let C_* be the cellular chain complex of the universal cover $\widetilde{M}$. Since M/ν is an homology 3-sphere and $S^3/Q(8)$ is the unique finite Swan complex for $Q(8)$, the complex $\mathbb{Z}[Q(8)] \otimes_{\mathbb{Z}[\pi]} C_*$ is $\mathbb{Z}[Q(8)]$-chain homotopy equivalent to the cellular chain complex for the universal cover of $S^3/Q(8)$. Any such chain homotopy equivalence may be realized by a $\mathbb{Z}[Q(8)]$-homology equivalence, since M and $S^3/Q(8)$ each have dimension $\leqslant 3$. However there are difficulties in carrying through our strategy (i.e., in extending Lemma 6.11) for this case, as $L_4(Q(8))$ has rank 5.

If $a > 1$ then $S^3/Q(8a)$ does not embed in any homology 4-sphere, by Corollary 4.4.2, and so we need not consider this possibility further, by Lemma 2.7.

The Sylow 2-subgroup of $Q(2^n a, b, c)$ is $Q(2^n)$, and is a retract of G. The inclusion induces isomorphisms on (co)homology with coefficients $\mathbb{F}_2$. If $n > 3$ and $M = Q(2^n)$ then restriction of ℓ_M to the 2-primary summand of τ_M is not hyperbolic, by Lemma 4.2. Since $\mathbb{Z}[\pi]$-homology equivalences preserve the linking forms, it follows from Lemma 2.7 that if $n > 3$ and π is an extension of $Q(2^n a, b, c)$ by a perfect normal subgroup then M does not embed in any homology 4-sphere.. Thus, the remaining possibilities are the groups $Q(8a, b, c)$, with $a > b > c \geqslant 1$. Subtle arithmetic arguments involving units in cyclotomic number fields have been used to show that certain groups of the form $Q(8p, q, 1)$ with p, q prime have non-trivial Swan invariant, and so do not act freely on homology spheres [**Mi83**]. However, these arguments do not exclude all such groups. We have seen no explicit examples, and it is not known whether any of the corresponding quotients could embed in S^4.

CHAPTER 7

The Complementary Regions

In this chapter we turn our attention to the variety of possible embeddings. We consider here $\chi(W)$ and $\pi_1 W$, for W a complementary region of an embedding of M in S^4. Our examples mostly involve Seifert manifolds M, and the obstructions to embeddings derive from the lower central series for π and its dual manifestation in terms of (Massey) products of classes in $H^1(M;\mathbb{Q})$.

We begin with a proof of the Generalized Schoenflies Theorem. This result is not specifically about embeddings in S^4, but the argument is so simple that it deserves a place here. The original proof of Aitchison's Theorem on smooth embeddings of $S^2 \times S^1$ used the Generalized Schoenflies Theorem. We sketch this briefly, and use surgery to give an argument for locally flat embeddings.

In the next three sections we use the Massey product structure in $H^*(M)$ to show that if M is a Seifert manifold with orientable base orbifold and non-zero Euler number, then $\chi(X) = \chi(Y) = 1$ is the only possibility. On the other hand, all values except for $\chi(X) = 1 - \beta$ and $\chi(Y) = 1 + \beta$ are realized by embeddings of $T_g \times S^1$.

Sections 7.6–7.8 lead to a criterion for a complementary region to be aspherical and of cohomological dimension at most 2. When $M = F \times S^1$ or when M is the total space of an S^1-bundle with non-orientable base, the simplest embeddings of M have one complementary component $X \simeq F$ and the other with cyclic fundamental group. In Section 7.9 we sketch how surgery may be used to identify such embeddings (up to s-cobordism). (No such argument is yet available when M fibres over an orientable base with Euler number 1.)

7.1. The Generalized Schoenflies Theorem

The arguments of Brown and Mazur for the Generalized Schoenflies Theorem each involve limiting processes in an essential way. Brown used ideas from decomposition theory ("Bing topology") involving collapsing cellular sets. The key idea in Mazur's argument is an ingenious regrouping of an infinite "sum", and we shall outline this version (as completed by M. Morse). Both versions are presented in [**Pu24**], and our account is based on this.

If M and N are two n-manifolds with fixed homeomorphisms $\partial M \cong \partial I^n$ and $\partial N \cong \partial I^n$, we let $M + N$ be the boundary connected sum, whereby we identify $(1, x_2, \ldots, x_n) \in \partial M$ with $(0, x_2, \ldots, x_n) \in \partial N$, for all $0 \leqslant x_2, \ldots, x_n \leqslant 1$. Clearly $M + D^n \cong M$, $M + N \cong N + M$, and stacking with respect to the first coordinate gives a natural identification of $\partial(M + N)$ with ∂I^n.

Define kM inductively by $1M = M$ and $(k+1)M = kM + M$, for $k \geqslant 1$. Then kM is embedded in $(k+1)M$, and $LM = \cup_{k\geqslant 1} kM$ is a (non-compact) n-manifold with boundary homeomorphic to $\mathbb{R}^{n-1}$. In particular, $LD^n \cong [0,\infty) \times I^{n-1} \cong \mathbb{R}^n_+$. We shall extend the definition of $+$ slightly by identifying $\{0\} \times I^{n-1} \subset \partial(1M) \subset LM$ with $\{1\} \times I^{n-1} \subset M$, and then we see that $M + LM \cong LM$. We also see that $L(M+N) \cong M + L(N+M)$.

THEOREM 7.1 (Brown-Mazur). *Let j be a locally flat embedding of S^{n-1} into S^n. Then there is a homeomorphism h of S^n such that $h \circ j$ is the equatorial embedding.*

PROOF. Let $E = S^{n-1} \times [-1,1]$. Since j is locally flat, it extends to an embedding $J : E \to S^n$ such that $J(s,0) = j(s)$, for all $s \in S^{n-1}$, by Brown's Collaring Theorem 2.1. Let X and Y be the complementary regions containing $J(S^{n-1} \times \{1\})$ and $J(S^{n-1} \times \{-1\})$, respectively, and fix homeomorphisms of ∂X and ∂Y with ∂D^n. We shall show that X and Y are n-discs.

After composition with a rotation of S^n, if necessary, we may assume that $J(P) = P$, for some $P \in S^{n-1} \times \{\frac{1}{2}\}$. Let $U = B_r(P) \subset J(S^{n-1} \times (0,1))$ be an open ball (with respect to the standard metric), and let $Q \in S^n \setminus X$ be another point. Note that $X \cong X \setminus (J(S^{n-1} \times (0,\frac{1}{2})) \cup U)$ and $Y \cong Y \cup J(S^{n-1} \times (0,\frac{1}{2})) \setminus U$, and so $X + Y \cong S^n \setminus U \cong D^n$.

There is a homeomorphism $c : S^n \setminus \{Q\} \cong U$, which is the identity on the closed ball $\overline{B_{r/2}(P)}$. Let $V = \overline{B_s(P)}$ be another closed ball centred on P such that $J(V) \subset \overline{B_{r/2}(P)}$. Let $W = J^{-1}(U)$, and let $f : S^n \setminus \{Q\} \to S^n$ be the composite of $(J|_W)^{-1} \circ c$ with the inclusion of U into S^n. The composite embedding $J' = f \circ J$ then has the property that there is an n-disc $D \subset S^{n-1} \times (0,1) \subset E$ such that the closed complements $\overline{S^n \setminus D}$ and $\overline{S^n \setminus J'(D)}$ are each homeomorphic to D^n. (Establishing this was the contribution of Morse.) It shall suffice to show that $f(X) \cong D^n$, since $X \subset S^n \setminus \{Q\}$ and f is a homeomorphism onto its image. We shall assume henceforth that the original embedding J has the above property.

Since $LD^n \cong \mathbb{R}^n_+$, the 1-point compactification $LD^n \cup \{\infty\}$ is homeomorphic to D^n. Now $X + L(Y+X) = L(X+Y) = LD^n$, and so $X \cong X + D^n \cong X + L(Y+X) \cup \{\infty\} \cong D^n$. A similar argument shows that $Y \cong D^n$.

Every self-homeomorphism ϕ of S^{n-1} extends to a self-homeomorphism Φ of D^n by setting $\Phi(r.s) = r\phi(s)$ for all $0 \leqslant r \leqslant 1$ and $s \in S^{n-1}$. We may extend j in this manner across each hemisphere to obtain a self-homeomorphism of S^n. The inverse homeomorphism h is the equatorial embedding. □

In this case there is no advantage in considering only 3-spheres in S^4. This result is also a consequence of TOP surgery, since the complementary regions must be contractible, by van Kampen's Theorem and Alexander duality. However, this is an unnecessarily blunt instrument here.

7.2. $S^2 \times S^1$ and the Aitchison Theorem

Since $S^2 \times S^1$ may be obtained by 0-framed surgery on the unknot, it has a standard abelian embedding with $X \cong S^1 \times D^3$ and $Y \cong D^2 \times S^2$. In fact $Y \cong D^2 \times S^2$

whenever $M = S^2 \times S^1$, by a result of Aitchison. This result predates topological surgery. The proof given in [**Rub80**] uses the "Dehn's Lemma" of R. A. Norman [**No69**] together with the Generalized Schoenflies Theorem to show that one complementary region of a *smooth* embedding of $S^2 \times S^1$ in S^4 must be homeomorphic to $S^2 \times D^2$. We shall outline this proof and then give one which applies to all (TOP locally flat) embeddings.

LEMMA (Norman). *Let D be an n-point smoothly immersed 2-disc in a 4-manifold P. Suppose that there is an embedded 2-sphere $\sigma \subset int P$ with trivial normal bundle such that $D \cap \sigma$ is a single point of transverse intersection. Then there is an embedded 2-disc $\Delta \subset P$ with $\partial\Delta = \partial P$.* □

If $j : S^2 \times S^1 \to S^4$ is an embedding then $\pi_Y = 1$, by Lemma 2.5, and so $C = j(\{*\} \times S^1)$ is null-homotopic in Y. Hence C bounds a smooth embedded disc Δ in Y, by Norman's Lemma. Let $N(\Delta)$ be a tubular neighbourhood of Δ in Y. Then $\overline{Y \setminus N(\Delta)}$ is homeomorphic to D^4, by the TOP Schoenflies Theorem, and so $Y \cong D^2 \times S^2$.

While the extension of transversality to the 4-dimensional TOP setting probably allows for a corresponding extension of the applicability of Norman's argument, we shall instead appeal to 1-connected TOP surgery.

THEOREM 7.2 (Aitchison). *If $M = S^2 \times S^1$ is embedded in S^4 then one complementary region is homeomorphic to $S^2 \times D^2$.*

PROOF. Let $j : M \to S^4$ be an embedding. Since $j_{X*} = \pi_1 j_X$ is a split monomorphism, $\pi_Y = 1$, by Lemma 2.5 and so $Y \simeq S^2$. Let $f : \partial Y \to S^2 \times S^1$ be a homeomorphism. Then f extends to a map from Y to $S^2 \times D^2$. Thus we obtain a map of pairs $F : (Y, \partial Y) \to (S^2 \times D^2, S^2 \times S^1)$ which is a homotopy equivalence and a homeomorphism on the boundary. Hence Y is homeomorphic to $S^2 \times D^2$, by 1-connected surgery [**FQ**, Theorem 11.6A]. □

If $\pi_X \cong \mathbb{Z}$ then $X \cong D^3 \times S^1$ and the embedding is equivalent to j_U, by Freedman's Unknotting Theorem for 2-knots [**FQ**, Theorem 11.7A]. (This is also a special case of Theorem 7.7.)

We shall give a partial extension of this result to embeddings of $\#^\beta(S^2 \times S^1)$ at the end of this chapter.

7.3. Massey Products

Massey products provide further obstructions to finding embeddings with given $\chi(X)$. For instance, if $H^2(X;\mathbb{Q}) \cong \mathbb{Q}$ or 0 then all triple Massey products $\langle a, b, c\rangle$ of elements $a, b, c \in H^1(X;\mathbb{Q})$ are proportional. Stallings's Theorem can be refined to relate "freeness" of quotients of the lower central series and its rational and *mod* (p) analogues to the vanishing of higher Massey products [**Dw75, St65**].

The $\mathbb{N}il^3$-manifold $M = M(Wh)$ has fundamental group $\pi \cong F(2)/\gamma_3 F(2)$, with a presentation

$$\pi = \langle x, y, z \mid z = xyx^{-1}y^{-1},\ xz = zx,\ yz = zy\rangle.$$

Every element of π has an unique normal form $x^m y^n z^p$. The images X, Y of x, y in $H_1(\pi) \cong H_1(T)$ form a (symplectic) basis. Let ξ, η be the Kronecker dual basis for $H^1(\pi)$. Define functions ϕ_ξ, ϕ_η and $\theta : \pi \to \mathbb{Z}$ by

$$\phi_\xi(x^m y^n z^p) = \frac{m(1-m)}{2}, \ \phi_\eta(x^m y^n z^p) = \frac{n(1-n)}{2} \text{ and } \theta(x^m y^n z^p) = -mn - p,$$

for all $x^m y^n z^p \in \pi$. (We consider these as inhomogeneous 1-cochains with values in the trivial π-module $\mathbb{Z}$.) Then

$$\delta\phi_\xi(g,h) = \xi(g)\xi(h), \quad \delta\phi_\eta(g,h) = \eta(g)\eta(h) \quad \text{and} \quad \delta\theta(g,h) = \xi(g)\eta(h),$$

for all $g, h \in \pi$. Thus $\xi^2 = \eta^2 = \xi \cup \eta = 0$, and the Massey triple products $\langle \xi, \xi, \eta \rangle$ and $\langle \xi, \eta, \eta \rangle$ are represented by the 2-cocycles $\phi_\xi \eta + \xi\theta$ and $\theta\eta + \xi\phi_\eta$, respectively. On restricting these to the subgroups generated by $\{x, z\}$ and $\{y, z\}$, we see that they are linearly independent.

In fact, $\langle \xi, \xi, \eta \rangle \cup \eta$ and $\langle \xi, \eta, \eta \rangle \cup \xi$ each generate $H^3(\pi; \mathbb{Q})$. This is best seen topologically. Let $p : M \to T$ be the natural fibration of M over the torus, and let x and y be simple closed curves in T which represent a basis for $\pi_1 \cong \mathbb{Z}^2$. The group $H_2(M) \cong \mathbb{Z}^2$ is generated by the images of fundamental classes of the tori $T_x = p^{-1}(x)$ and $T_y = p^{-1}(y)$. If we fix sections in M for the loops x and y we see that $[T_x] \bullet x = [T_y] \bullet y = 0$ while $|[T_x \bullet y| = |T_y \bullet x| = 1$. Hence $[T_x]$ and $[T_y]$ are Poincaré dual to η and ξ, respectively. Since $\langle \xi, \xi, \eta \rangle$ restricts non-trivially to T_x and trivially to T_y we must have $\langle \xi, \xi, \eta \rangle \cup \eta \neq 0$, and similarly $\langle \xi, \eta, \eta \rangle \cup \xi \neq 0$. Similarly, $\langle \xi, \xi, \eta \rangle \cup \xi = \langle \xi, \eta, \eta \rangle \cup \eta = 0$. Thus, these Massey products are the Poincaré duals of Y and X, respectively.

Since the components of Wh are unknotted, M embeds in S^4, with $\chi(X) = \chi(Y) = 1$, and $\mu_M = 0$, since $\beta = 2$. On the other hand, M has no embedding with $\chi(X) = -1$, for otherwise $H^3(X)$ would contain $\langle \xi, \xi, \eta \rangle \cup \eta$, and so be non-trivial.

A similar strategy may be used for $M = M(g; (1, e))$ and $\pi = \pi_1 M$, when $g > 1$.

LEMMA 7.3. *Let $p : M \to T_g$ be the projection of an S^1-bundle with non-zero Euler Number. Then $H^1(p; \mathbb{Q})$ is an isomorphism and $H^2(p; \mathbb{Q}) = 0$, and so $\mu_M = 0$. Let $\{\alpha_1, \beta_1, \dots, \alpha_g, \beta_g\}$ be the basis for $H = H^1(M; \mathbb{Q})$ which is Kronecker dual to a symplectic basis for $H_1(M; \mathbb{Q}) \cong H_1(T_g; \mathbb{Q})$. Then the Massey triple products $\langle \alpha_i, \alpha_i, \beta_i \rangle$ and $\langle \alpha_i, \beta_i, \beta_i \rangle$ (for $1 \leqslant i \leqslant g$) form a basis for $H^2(\pi; \mathbb{Q})$. This basis is Poincaré dual to the given basis for $H_1(\pi; \mathbb{Q})$.*

PROOF. The first assertion follows from the Gysin sequence for the bundle [**Span**, Theorem 5.7.11]. We may assume that $\pi = \pi_1 M$ has a presentation $\langle x_1, y_1, \dots, x_g, y_g \mid \Pi[x_i, y_i] = h^e, \ h \ \textit{central} \rangle$. Thus, the images of the generators $\{x_1, y_1, \dots, x_g, y_g\}$ in $H_1(B)$, represent a standard symplectic basis, and so determine a basis for $H_1(M; \mathbb{Q}) \cong H_1(B; \mathbb{Q})$. The argument then follows as in the case of $M(1, (1, -1))$ discussed above. □

We shall extend these results to the Seifert case in the next section.

7.4. Seifert Manifolds

In this section we shall use cup products and Massey products to restrict the possible values of $\chi(W)$ for W a complementary region of an embedding of a Seifert manifold M with orientable base orbifold.

LEMMA 7.4. **[CT14]** *Let M be a Seifert manifold. If the base B is non-orientable or if $\varepsilon(M) \neq 0$ then $H^*(M;\mathbb{Q}) \cong H^*(\#^\beta S^2 \times S^1;\mathbb{Q})$. Otherwise, the image of h in $H_1(M;\mathbb{Q})$ is non-zero, and $H^*(M;\mathbb{Q}) \cong H^*(|B| \times S^1;\mathbb{Q})$.*

PROOF. There is a finite regular covering $q : \widehat{M} \to M$, where $\widehat{M}$ is an S^1-bundle space with orientable base $\widehat{B}$, say. Let $G = Aut(q)$. Then $H^*(M;\mathbb{Q}) \cong H^*(\widehat{M};\mathbb{Q})^G$. If B is non-orientable or if $\varepsilon(M) \neq 0$ then the regular fibre has image 0 in $H_1(M;\mathbb{Q})$, and so $H^*(\widehat{B};\mathbb{Q})$ maps onto $H^*(\widehat{M};\mathbb{Q})$. Hence all triple cup products of classes in $H^1(\widehat{M};\mathbb{Q})$ are 0. Therefore, all pairwise cup products of such classes are also 0, by the non-degeneracy of Poincaré duality, and so $H^*(M;\mathbb{Q}) \cong H^*(\#^\beta S^2 \times S^1;\mathbb{Q})$. Otherwise, $\widehat{M} \cong \widehat{B} \times S^1$ and G acts orientably on each of S^1 and $\widehat{B}$. Hence the image of h in $H_1(M;\mathbb{Q})$ is non-zero and $H^*(M;\mathbb{Q}) \cong H^*(|B| \times S^1;\mathbb{Q})$. □

We may use the observations on cup product from Lemma 7.4 to extract some information on the image of the regular fibre under the maps $H_1(j_X)$ and $H_1(j_Y)$, when M is Seifert fibred.

THEOREM 7.5. *Let $M = M(g;S)$ where $g \geqslant 1$ and $\varepsilon(M) = 0$. If M embeds in S^4 then $\chi(X) > 1 - \beta = -2g$ and $\chi(Y) < 1 + \beta = 2g + 2$. If $\chi(X) < 0$ then the image of h in $H_1(Y;\mathbb{Q})$ is non-trivial.*

PROOF. Let $\{a_i^*, b_i^*; 1 \leqslant i \leqslant g\}$ be the images in $H^1(M;\mathbb{Q})$ of a symplectic basis for $H^1(|B|;\mathbb{Q})$. Then $a_i^*(h) = b_i^*(h) = 0$ for all i. Let $\theta \in H^1(M;\mathbb{Q})$ be such that $\theta(h) \neq 0$. By Lemma 7.4 we have

$$H^*(M;\mathbb{Q}) \cong H^*(|B| \times S^1;\mathbb{Q}) \cong \mathbb{Q}[\theta, a_i^*, b_i^*, \ \forall\, i \leqslant g]/I,$$

where I is the ideal $(\theta^2, a_i^{*2}, b_i^{*2}, \theta a_i^* b_i^* - \theta a_j^* b_j^*, a_i^* a_j^*, b_i^* b_j^*, \ \forall\, 1 \leqslant i < j \leqslant g)$.

Since $\theta a_1^* b_1^* \neq 0$, the triple product $\mu_M \neq 0$, and so M has no embedding with $\beta_2(Y) = 0$, by Lemma 2.3. Hence $\chi(X) = 1 - \beta$ ($\Leftrightarrow \chi(Y) = 1 + \beta$) is impossible.

If $\chi(X) < 0$ then $\beta_1(X) > g + 1$, and so the image of $H^1(X;\mathbb{Q})$ in $H^1(M;\mathbb{Q})$ must contain some pair of classes from the image of $H^1(|B|;\mathbb{Q})$ with non-zero product. But then it cannot also contain θ, since all triple products of classes in $H^1(X;\mathbb{Q})$ are 0. Thus, the image of $H^1(Y;\mathbb{Q})$ must contain a class which is non-trivial on h, and so $j_Y(h) \neq 0$ in $H_1(Y;\mathbb{Q})$. □

In particular, if $g = 1$ then $\chi(X) = 0$ and $\chi(Y) = 2$.

Theorem 7.5 also follows from Lemma 2.4, since the centre of π is not contained in the commutator subgroup π'.

If the base orbifold B is non-orientable or if $\varepsilon(M) \neq 0$ then $\mu_M = 0$, by Lemma 7.4, and so the argument of Theorem 7.5 does not extend to these cases. However, Lemma 7.4 also suggests that when $\varepsilon(M) \neq 0$ we should be able to use Massey product arguments as in Section 6.3 (where we considered the case $S = \emptyset$).

THEOREM 7.6. *Let $M = M(g; S)$, where $g \geqslant 0$ and $\varepsilon(M) \neq 0$. If M embeds in S^4 with complementary regions X and Y then $\chi(X) = \chi(Y) = 1$.*

PROOF. The group $\pi = \pi_1 M(g; S)$ has a presentation

$$\langle x_1, y_1, \dots, x_g, y_g, c_1, \dots, c_r, h \mid \Pi[a_i, b_i]\Pi c_j = 1,\ c_i^{\alpha_i} h^{\beta_i} = 1,\ h \text{ central}\rangle.$$

We may assume that $g \geqslant 1$, for if $g = 0$ then M is a $\mathbb{Q}$-homology 3-sphere and the result is clear. As in Lemma 7.4, there is a finite regular covering $q : \widehat{M} \to M$, where $\widehat{M}$ is an S^1-bundle space with orientable base $\widehat{B}$. Let $G = Aut(q)$. Then $H^*(M; \mathbb{Q}) \cong H^*(\widehat{M}; \mathbb{Q})^G$. Let $\{\alpha_2, \beta_2, \dots, \alpha_h, \beta_h\}$ be a basis for $H^1(\widehat{M}; \mathbb{Q})$ which is Kronecker dual to the image of a symplectic basis for $H_1(\widehat{B})$ in $H_1(\widehat{M}; \mathbb{Q})$.

The homomorphism $H^1(p; \mathbb{Q})$ from $H^1(|B|; \mathbb{Q})$ to $H^1(M; \mathbb{Q})$ induced by the Seifert fibration $p : M \to B$ is an isomorphism, by Lemma 7.4. If $j : M \to S^4$ is an embedding then $H = H^1(M; \mathbb{Q}) = A \oplus \Omega$, where A and Ω are self-annihilating with respect to cup product. If $L \leqslant H$ is a direct summand of rank $> g$ then there are $a \in L \cap A$ and $b \in L \setminus A$ such that $a \cup b \neq 0$ in $H^2(|B|; \mathbb{Q})$. We consider their images under $H^1(q; \mathbb{Q})$. Since $\widehat{M} \cong \widehat{B}_o \times S^1 \cup D^2 \times S^1$, it is easy to see that every self-homeomorphism of $\widehat{B}$ lifts to a fibre-preserving self-homeomorphism of $\widehat{M}$. Thus (after multiplication by factors in $\mathbb{Q}^\times$, if necessary) we may assume that $a = \alpha_1$ and then $b = \beta_1 + b'$, where b' is in the span of $\{\alpha_2, \beta_2, \dots, \alpha_h, \beta_h\}$. We now view these classes as classes in $H^1(\widehat{B}; \mathbb{Q})$. Since $a \cup b' = 0$ there is a map $f : \widehat{B} \to V = S^1 \vee S^1$, such that a and b' are in the image of $H^1(f; \mathbb{Q})$. Hence $\langle a, a, b' \rangle = 0$. The topological argument used in Section 6.2 shows that $\langle a, a, \beta_1 \rangle \cup b' = 0$ also. Therefore, $\langle a, a, b \rangle \cup b = \langle \alpha_1, \alpha_1, \beta_1 \rangle \cup \beta_1$ is non-zero. But this contradicts the fact that $H^3(X) = H^3(Y) = 0$. Therefore, $H^1(X)$ and $H^1(Y)$ each have rank at most g, and so $\chi(X) = \chi(Y) = 1$. □

If $\chi(X) = 0$, all cone point orders are odd and h has non-zero image in $H_1(X; \mathbb{Q})$ then S is skew-symmetric, by Theorem 3.10.1. (In particular, this must be the case if g and ε_S are 0.) Conversely, if S is skew-symmetric and all cone point orders a_i are odd then $M(0; S)$ embeds smoothly. Since $\beta = 1$ we must have $\chi(X) = 0$ and $H_1(Y; \mathbb{Q}) = 0$. (In fact, for the embedding constructed on page 693 of [**CH98**] the component X has a fixed-point-free S^1-action.) Hence also $M(g; S)$ embeds smoothly, as in Lemma 3.8, which gives embeddings with $\chi(X) = 0$. Is there a natural choice of 0-framed bipartedly slice link representing $M(0; S)$? Are all values of $\chi(X)$ consistent with Theorem 7.5 possible for $M(g; S)$?

However, even if $\chi(X) = 0$ the other hypothesis of Theorem 3.10.1 need not hold. For instance, the standard 0-framed link representing $M = T_2 \times S^1$ has two essentially different partitions into 3- and 2-component trivial sublinks. For one, $\pi_X \cong \mathbb{Z} \times F(2)$ and $\pi_Y \cong F(2)$, while for the other $\pi_X \cong \mathbb{Z} * \mathbb{Z}^2$ and $\pi_Y \cong \mathbb{Z}^2$.

If ℓ_M is hyperbolic then all even cone point orders have the same 2-adic valuation, by Theorem 4.3 (when $g < 0$) and Lemma 6 of Appendix A (when $g \geqslant 0$).

7.5. S^1-Bundle Spaces

The bundle space $E = M(g;(1,e))$ can only embed in S^4 if $e = 0$ or ± 1, since $\tau_E = 0$ if $e = 0$ and is cyclic of order e otherwise. The 3-torus $M(1;(1,0))$ is also $M(Bo)$. Since $M(g;(1,0)) \cong T_g \times S^1$ is an iterated fibre sum of copies of $T \times S^1$, it may be obtained by 0-framed surgery on the $(2g+1)$-component sublink of the link of Figure 3.4 consisting of the central loop and the loops along the top of this figure.

This 3-manifold has an embedding as the boundary of $T_g \times D^2$, the regular neighbourhood of the unknotted embedding of T_g in S^4, with the other complementary region having fundamental group $\mathbb{Z}$. It is easy to see that if $g \geqslant 1$ then $T_g \times S^1$ has other embeddings with $\chi(X)$ realizing each even value $> 1 - \beta$. On the other hand, $\mu_{T_g \times S^1} \neq 0$, and so no embedding has a complementary region Y with $\beta_1(Y) = 0$.

Changing the framing on one component of Bo to 1, and applying a Kirby move to isolate this component gives the disjoint union of the Whitehead link Wh and the unknot. Since the linking numbers are 0 the framings are unchanged, and we may delete the isolated 1-framed unknot. Thus $M(1;(1,1))$ may be obtained by 0-framed surgery on Wh. The corresponding modification of the standard 0-framed $(2g+1)$-component link L representing $T_g \times S^1$ involves changing the framing of the component L_{2g+1} whose meridian represents the central factor of π. Performing a Kirby move and deleting an isolated 1-framed unknot gives a 0-framed $2g$-component link representing $M(g;(1,1))$.

Since the original link had partitions into two trivial links, with $g+1$ and g components, respectively, the new link has a partition into two trivial g-component links. However, this is the only partition into slice sublinks. As we shall see, consideration of the Massey product structure shows that all embeddings of $M(g;(1,1))$ have $\chi(X) = \chi(Y) = 1$.

Suppose now that F is non-orientable. Then $M(-c;(1,e))$ embeds if and only if it embeds as the boundary of a regular neighbourhood of an embedding of $\#^c\mathbb{RP}^2$ with normal Euler number e. We must have $e \leqslant 2c$ and $e \equiv 2c \; mod \; (4)$, by Corollary 4.4.2. The standard embedding of $\mathbb{RP}^2$ in S^4 is determined up to composition with a reflection of S^4. The complementary regions are each homeomorphic to a disc bundle over $\mathbb{RP}^2$ with normal Euler number 2, and so have fundamental group $\mathbb{Z}/2\mathbb{Z}$. The standard embeddings of $\#^c\mathbb{RP}^2$ are obtained by taking iterated connected sums of these building blocks $\pm(S^4, \mathbb{RP}^2)$, and in each case the exterior has fundamental group $\mathbb{Z}/2\mathbb{Z}$. The regular neighbourhoods of $\#^c\mathbb{RP}^2$ are disc bundles with boundary $M(-c;(1,e))$. Thus $M(-c;(1,e))$ has an embedding with one complementary component $X_{c,e}$ a disc bundle over $\#^c\mathbb{RP}^2$ and the other component $Y_{c,e}$ having fundamental group $\mathbb{Z}/2\mathbb{Z}$.

This embedding arises from a 0-framed $(c+1)$-component link assembled from copies of the $(2,4)$-torus link 4_1^2 and its reflection. This is the union of an unknot and a trivial c-component link, but has no other partitions into slice links. However, we can do better if we recall that $\#^c\mathbb{RP}^2 \cong (\#^{c-2g}\mathbb{RP}^2)\#T_g$ for any g such that $2g < c$. The 3-manifold obtained by 0-framed surgery on the $(a+b+2g+1)$-component link of Figure 7.1 is $M(-c;(1,e))$, where $c = a + b + 2g$ and $e = \pm 2(a - b)$, with the sign depending on the choice of orientation. (See also [**CH98**, Figure A.3].)

This link has partitions into trivial sublinks corresponding to all the values $2 - c \leqslant \chi(X) \leqslant \min\{2 - \frac{|e|}{2}, 1\}$ such that $\chi(X) \equiv c \ mod\ (2)$. Are any other values realized? In particular, does $M(-3;(1,6))$ embed with $\chi(X) = \chi(Y) = 1$?

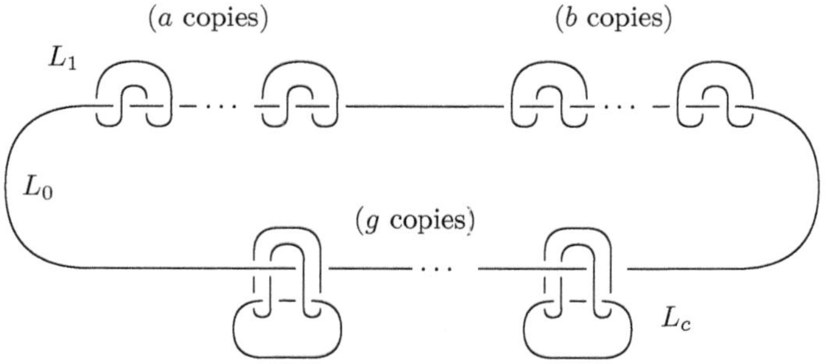

Figure 7.1 A link giving $M(-c;(1,e))$.

If we move beyond the class of S^1-bundle spaces, we may give an example of "intermediate" behaviour. It is not hard to show that if $H \cong \mathbb{Z}^\beta$ with $\beta \leqslant 5$ then for every $\mu : \wedge^3 H \to \mathbb{Z}$ there is an epimorphism $\lambda : H \to \mathbb{Z}$ such that μ is 0 on the image of $\wedge^3\mathrm{Ker}(\lambda)$. Hence there are splittings $H \cong A \oplus B$ with A of rank 3 or 4 such that μ restricts to 0 on each of $\wedge^3 A$ and $\wedge^3 B$. However if $\beta = 6$ this fails for

$$\mu = e_1 \wedge e_2 \wedge e_3 + e_1 \wedge e_5 \wedge e_6 + e_2 \wedge e_4 \wedge e_5.$$

(Here $\{e_i\}$ is the basis for $Hom(H,\mathbb{Z})$ which is Kronecker dual to the standard basis of $H \cong \mathbb{Z}^6$.) For every epimorphism $\lambda : \mathbb{Z}^6 \to \mathbb{Z}$ there is a rank 3 direct summand A of $\mathrm{Ker}(\lambda)$ such that μ is non-trivial on $\wedge^3 A$. [This requires a little calculation. Suppose that $\lambda = \Sigma\lambda_i e_i^*$. If $\lambda_6 \neq 0$, then we let $f_j = \lambda_6 e_j - \lambda_j e_6$, for $1 \leqslant j \leqslant 3$. Then $\mu(f_1 \wedge f_2 \wedge f_3) = \lambda_6^3 \neq 0$, and so we may take A to be the direct summand containing $\langle f_1, f_2, f_3\rangle$. A similar choice applies if λ_3 or λ_4 is non-zero. If $\lambda_3 = \lambda_4 = \lambda_6 = 0$ but $\lambda_1 \neq 0$ then we may take A to be the direct summand containing $\langle g_2, e_4, g_5\rangle$, where $g_2 = \lambda_1 e_2 - \lambda_2 e_1$ and $g_5 = \lambda_1 e_5 - \lambda_5 e_1$. A similar choice applies if λ_2 or λ_5 is non-zero.]

This example arose in a somewhat different context. It is the cup product 3-form of the 3-manifold M given by 0-framed surgery on the 6-component link of [**DH17**, Figure 6.1]. This link has certain "Brunnian" properties. All the 2-component sublinks, all but three of the 3-component sublinks and six of the 4-component sublinks are trivial. Thus, M has embeddings in S^4 with $\chi(X) = -1$ or 1, corresponding to partitions of L into a pair of trivial sublinks, but there are no embeddings with $\chi(X) = -5$ or -3, since μ_M does not satisfy the second assertion of Lemma 2.

7.6. Homotopy Types of Pairs

We shall next give some lemmas on recognizing the homotopy types of certain spaces and pairs of spaces arising later.

We shall say that $c.d.W \leqslant n$ if the equivariant chain complex of the universal cover $\widetilde{W}$ is chain homotopy equivalent to a complex of projective $\mathbb{Z}[\pi_1 W]$-modules of length $\leqslant n$. (If W is aspherical then $c.d.W = c.d.\pi_1 W$, as defined in [**Bro**].)

THEOREM 7.7. *Let U and V be connected finite cell complexes such that $c.d.U \leqslant 2$ and $c.d.V \leqslant 2$. If $f : U \to V$ is a 2-connected map then $\chi(U) \geqslant \chi(V)$, with equality if and only if f is a homotopy equivalence.*

PROOF. Up to homotopy, we may assume that f is a cellular inclusion, and that V has dimension $\leqslant 3$. Let $\pi = \pi_1 U$ and let $C_* = C_*(\widetilde{V}, \widetilde{U})$. Then $H_q(C_*) = 0$ if $q \leqslant 2$, since f is 2-connected, and $H_q(C_*) = 0$ if $q > 3$, since $c.d.U$ and $c.d.V \leqslant 2$. Hence $H_3(C_*) \oplus C_2 \oplus C_0 \cong C_3 \oplus C_1$, by Schanuel's Lemma, and so $H_3(C_*)$ is a stably free $\mathbb{Z}[\pi]$-module of rank $-\chi(C_*) = \chi(U) - \chi(V)$. Hence $\chi(U) \geqslant \chi(V)$, with equality if and only if $H_3(C_*) = 0$, since group rings are weakly finite, by a theorem of Kaplansky. (See [**Ros84**] for a proof of Kaplansky's result.) The result follows from the long exact sequence of the pair $(\widetilde{Y}, \widetilde{X})$ and the theorems of Hurewicz and Whitehead. □

If X is a finite complex and $c.d.X \leqslant 2$ then $C_*(\widetilde{X})$ is chain homotopy equivalent to a finite projective complex of length 2, which is a partial resolution of the augmentation module $\mathbb{Z}$. Chain homotopy classes of such partial resolutions are classified by $Ext^3_{\mathbb{Z}[\pi]}(\mathbb{Z}, \Pi) = H^3(\pi; \Pi)$, where Π is the module of 2-cycles.

COROLLARY 7.7.1. *If U is a connected finite complex such that $c.d.U \leqslant 2$ and $\pi_1 U \cong \mathbb{Z}$ then $U \simeq S^1 \vee \bigvee^{\chi(U)} S^2$.*

PROOF. Since $c.d.U \leqslant 2$ and projective $\mathbb{Z}[\pi_1 U]$-modules are free, $C_*(\widetilde{U})$ is chain homotopy equivalent to a finite free $\mathbb{Z}[\pi_1 U]$-complex P_* of length $\leqslant 2$, and $\chi(U) = \Sigma(-1)^i rank(P_i)$. Since $\pi_2 U \cong H_2(U; \mathbb{Z}[\pi_1 U])$ is the module of 2-cycles in $C_*(\widetilde{U})$, it is free of rank $\chi(U)$. Let $f : S^1 \vee \bigvee^{\chi(U)} S^2 \to U$ be the map determined by a generator for $\pi_1 U$ and representatives of a basis for $\pi_2 U$. Then f is a homotopy equivalence, by Theorem 7.7. □

Theorem 3.2 of [**FMGK**] gives an analogue of Theorem 7.7 for maps between closed 4-manifolds. The argument extends to the following relative version.

LEMMA 7.8. *Let $f : (X_1, A_1) \to (X_2, A_2)$ be a map of orientable PD_4-pairs such that $f|_{A_1} : A_1 \to A_2$ is a homotopy equivalence. Then f is a homotopy equivalence of pairs if and only if $\pi_1 f$ is an isomorphism and $\chi(X_1) = \chi(X_2)$.*

PROOF. Since $f|_{A_1} : A_1 \to A_2$ is a homotopy equivalence, f has degree 1, and hence is 2-connected as a map from X_1 to X_2. The rest of the argument is as in [**FMGK**, Theorem 2]. □

In certain cases we can identify the homotopy type of a pair.

LEMMA 7.9. *Let (X, A) and (X', A') be pairs such that the inclusions $\iota_A : A \to X$ and $\iota_{A'} : A' \to X'$ induce epimorphisms on fundamental groups. If X and X' are aspherical and $f : A \to A'$ is a homotopy equivalence such that $\pi_1 f(\mathrm{Ker}(\pi_1\iota_A)) = \mathrm{Ker}(\pi_1\iota_{A'})$ then f extends to a homotopy equivalence of pairs $(X, A) \simeq (X', A')$.*

PROOF. The fundamental group conditions imply that $g = \iota_{A'}f$ extends to a map from the relative 2-skeleton $X^{[2]} \cup A$. The further obstructions to extending g to a map from X to X' lie in $H^{q+1}(X, A; \pi_q(X'))$, for $q \geqslant 2$. Since X' is aspherical these groups are 0. The other hypotheses imply that any extension $h : X \to X'$ induces an isomorphism on fundamental groups, and hence is a homotopy equivalence. □

7.7. Cohomological Dimension and Fundamental Group

Since the complementary regions are 4-manifolds with non-empty boundary they are homotopy equivalent to 3-dimensional complexes. In general, it remains an open question whether such a space must be homotopically 2-dimensional. However, there is a simple criterion for the complementary regions to have cohomological dimension 2, meaning that all homology and cohomology groups with arbitrary local coefficients are trivial in degrees > 2.

In the next theorem we refer to the Bass Conjectures [**Ba76**], which we outline very briefly. If P is a finitely generated projective $\mathbb{Z}[\pi]$-module then it is the image of an idempotent $n \times n$-matrix A with entries in $\mathbb{Z}[\pi]$, for some $n \geqslant 0$. The Kaplansky rank $\kappa(P)$ is the coefficient of 1 in the trace of A. The weak Bass Conjecture is the assertion that $\kappa(P)$ equals the "naive" rank $\dim_{\mathbb{Q}} \mathbb{Q} \otimes_{\mathbb{Z}[\pi]} P$. An analytic argument originally due to Kaplansky shows that $\kappa(P) > 0$ if $P \neq 0$.

THEOREM 7.10. *Let W be a complementary region of an embedding of M in S^4. Then c.d.$W \leqslant 2$ if and only if $j_{W*} = \pi_1 j_W$ is an epimorphism. If so, then W is aspherical if and only if c.d.$\pi_1 W \leqslant 2$ and $\chi(W) = \chi(\pi_1 W)$.*

PROOF. Let $\Gamma = \mathbb{Z}[\pi_1 W]$. Then there are Poincaré-Lefshetz duality isomorphisms $H_i(W; \Gamma) \cong H^{4-i}(W, \partial W; \Gamma)$ and $H^j(W; \Gamma) \cong H_{4-j}(W, \partial W; \Gamma)$, for all $i, j \leqslant 4$.

If *c.d.*$W \leqslant 2$ then $H_i(\widetilde{W}, \partial\widetilde{W}) = H_i(W, \partial W; \Gamma) = 0$ for $i \leqslant 1$, and so $\partial\widetilde{W}$ is connected. Therefore j_{W*} must be surjective. Conversely, if j_{W*} is an epimorphism, then we may assume that W may be obtained from $M = \partial W$ (up to homotopy) by adjoining cells of dimension $\geqslant 2$. Hence $H_i(W, \partial W; \Gamma)$ and $H^j(W, \partial W; \Gamma)$ are 0 for $i, j \leqslant 1$. Therefore $H_q(W; \Gamma) = H^q(W; \Gamma) = 0$ for all $q > 2$, and so $C_*(W; \Gamma)$ is chain homotopy equivalent to a complex P_* of finitely generated projective Γ-modules of length at most 2, by Wall's finiteness criteria [**Wa66**].

If W is aspherical then *c.d.*$\pi_1 W \leqslant 2$, and we must have $\chi(W) = \chi(\pi_1 W)$. Conversely, if j_{W*} is an epimorphism, then $\Pi = H_2(P_*) \cong \pi_2 W$ is the only obstruction to asphericity. If, moreover, *c.d.*$\pi_1 W \leqslant 2$ we may apply Schanuel's Lemma, to see that P_* splits as

$$P_* = \Pi \oplus (Z_1 \to P_1 \to P_0),$$

where π is concentrated in degree 2, Z_1 is the submodule of 1-cycles and $Z_1 \to P_1 \to P_0$ is a resolution of the augmentation module $\mathbb{Z} = H_0(P_*)$. Now $\mathbb{Z} \otimes_\Gamma \Pi \cong H_2(W)$ is a free abelian group of rank $\chi(W) - \chi(\pi_1 W)$. If, moreover, $\chi(W) = \chi(\pi_1 W)$ then $\Pi = 0$, and so W is aspherical, since the weak Bass Conjecture holds for groups of cohomological dimension $\leqslant 2$ [**Ec01**]. □

COROLLARY 7.10.1. *The augmentation ideal of the group ring* $\mathbb{Z}[\pi_X]$ *has a square presentation matrix.*

PROOF. If $c.d.X \leqslant 2$ then $C_*(X; \mathbb{Z}[\pi])$ is chain homotopy equivalent to a finite free $\mathbb{Z}[\pi_X]$-complex of length 2. Hence the augmentation ideal of the group ring $\mathbb{Z}[\pi_X]$ has a square presentation matrix, since $\chi(X) \leqslant 1$. □

The property that the augmentation ideal has a square presentation matrix interpolates between π_X having a balanced presentation and being homologically balanced. The stronger condition (having a balanced presentation) would hold if X were homotopy equivalent to a finite 2-dimensional cell complex.

In our applications of Theorem 7.10, $\pi_1 W$ is either free, free abelian or the fundamental group of an aspherical surface. Hence all projective Γ-modules are stably free. A stably free Γ-module P is trivial if and only if $\mathbb{Z} \otimes_\Gamma P = 0$, by an old result of Kaplansky (see [**Ros84**] for a proof), and we could use this instead of invoking [**Ec01**]. A similar argument may be used to show that, in general, W is aspherical if and only if $c.d.\pi_1 W \leqslant 3$, $\pi_1 W$ is of type FF, $\chi(W) = \chi(\pi_1 W)$ and $\pi_2 W = 0$.

Let K be the Artin spin of a non-trivial classical knot, and let $X = X(K)$ be the exterior of a tubular neighbourhood of K in S^4. Then $\pi_1 X \cong \pi K$, the knot group, and $M = \partial X \cong S^2 \times S^1$. In this case $c.d.\pi K = 2$ and $\chi(X) = \chi(\pi K) = 0$, but j_{X*} is not onto, and X is not aspherical. (Thus $c.d.X = 3$.)

There are two essentially different partitions of the standard link representing $T_g \times S^1$ into moieties with $g+1$ and g components. For one, $X \cong S^1 \times (\natural^g (D^2 \times S^1)$, which is aspherical (as to be expected from Theorem 7.10); for the other, $\pi_X \cong \mathbb{Z}^2 * F(g-1)$, and X is not aspherical. (In neither case is Y aspherical.)

The following lemma is inspired by the arguments of [**FMGK**, Chapter 2.3], but here we do not assume that X is homotopy equivalent to a finite 2-complex. (See [**Lück**] for details of the von Neumann algebra $\mathcal{N}(\pi_X)$ invoked here.)

LEMMA 7.11. *If* $c.d.X \leqslant 2$ *and* $\beta_1^{(2)}(\pi_X) = 0$ *then either* $\chi(X) = 0$ *and* X *is aspherical or* $\chi(X) > 0$. *Hence if* π_X *is elementary amenable and* $\chi(X) = 0$ *then* $\pi_X \cong \mathbb{Z}$ *or* $BS(1,m)$ *for some* $m \neq 0$.

PROOF. This follows from a mild extension of [**FMGK**, Theorem 2.5]. Since $c.d.X \leqslant 2$ and X is homotopy equivalent to a finite 3-complex, $C_*(\widetilde{X})$ is chain homotopy equivalent to a finite free $\mathbb{Z}[\pi_X]$-complex D_* of length at most 2. If $\beta_1^{(2)}(\pi_X) = 0$ then π_X is infinite, so $H_i(\mathcal{N}(\pi_X) \otimes_{\mathbb{Z}[\pi]} D_*) = 0$ for $i \leqslant 1$. Hence $\chi(X) = \chi(D_*) = dim_{\mathcal{N}(\pi_X)} H_2(\mathcal{N}(\pi_X) \otimes_{\mathbb{Z}[\pi]} D_*) \geqslant 0$, with equality only if D_* is acyclic, in which case X is aspherical.

The second assertion follows from [**FMGK**, Corollary 2.6.1]. □

EXAMPLE 7.12. $X \simeq Kb$.
Let $M = M(-2;(1,0))$ or $M(-2;(1,4))$ and suppose that j is bi-epic. In each case π is polycyclic and $\pi/\pi' \cong \mathbb{Z}\oplus(\mathbb{Z}/2\mathbb{Z})^2$. Hence $\chi(X) = 0$, and so $c.d.\pi_X \leqslant 2$. Since π_X is a quotient of π and $\pi_X/\pi'_X \cong \mathbb{Z}\oplus\mathbb{Z}/2\mathbb{Z}$ we must have $\pi_X \cong \mathbb{Z}\rtimes_{-1}\mathbb{Z}$. Since $c.d.X \leqslant 2$ and $\chi(X) = 0$ the classifying map $c_X : X \to Kb = K(\mathbb{Z}\rtimes_{-1}\mathbb{Z}, 1)$ is a homotopy equivalence.

7.8. Aspherical Embeddings

An embedding j is *aspherical* if each of the complementary regions is apherical. Every homology sphere has an aspherical embedding, since it bounds a contractible 4-manifold. Sums of aspherical embeddings are again aspherical.

Suppose that a complementary region W is aspherical. Then $c.d.\pi_W \leqslant 3$, and π_W cannot be a PD_3-group since $H^3(W) = 0$. If $c.d.\pi_W \leqslant 2$ then j_{W*} is an epimorphism, by equivariant Poincaré duality [**DH25**]. Hence if j is aspherical then *j is bi-epic $\Leftrightarrow$ $c.d.\pi_X \leqslant 2$ and $c.d.\pi_Y \leqslant 2$*, by Theorem 7.10.

If G and H are two groups of cohomological dimension 2 with balanced presentations and isomorphic abelianizations then there is an embedding of a 3-manifold such that the complementary regions X and Y are each homotopy equivalent to finite 2-complexes and $\pi_X \cong G$ and $\pi_Y \cong H$ [**Lic04**]. Moreover $\chi(X) = \chi(\pi_X) = 1$ and $\chi(Y) = \chi(\pi_Y) = 1$. If, moreover, $\beta_2(G) = \beta_1(G)$ and $\beta_2(H) = \beta_1(H)$ then X and Y are aspherical [**FMGK**, Theorem 2.8].

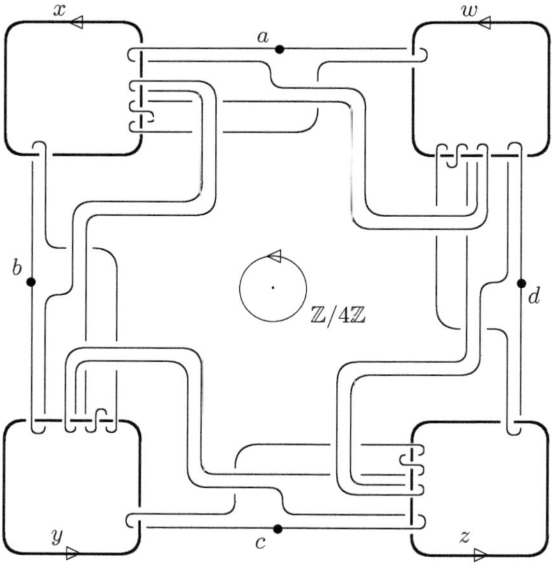

Figure 7.2 $X \cong Y \simeq K(Hig, 1)$.

EXAMPLE 7.13. *Generalized Higman groups.*
The simplest examples of this type we know of are based on the groups with presentations of the form

$$\langle a,b,c,d \mid a^m = bab^{-1},\ b^m = cbc^{-1},\ c^m = dcd^{-1},\ d^m = ada^{-1}\rangle.$$

We may construct such groups by assembling copies of $BS(1,m)$ over free subgroups. (See [**Rob**, Exercise 6.4.15].) Hence they all have cohomological dimension 2. In particular, $m=1$ gives the product $F(2)\times F(2)$, while $m=2$ gives the Higman group Hig, with $H_1(Hig) = Hig^{ab} = 1$. Since the presentations are balanced, it follows that $H_2(Hig)=0$, and so Hig is superperfect.

Figure 7.2 is symmetric under quarter-turn rotations around the axis through the central point.

Consider the embedding corresponding to the link in Figure 7.3, in which the strands in the box have m full twists, and the central component represents the word $A = uvu^{-1}v^{-m}$ in the meridians u,v for the other components. When $m=0$ the link is the split union of an unknot and the Hopf link, $M\cong S^2\times S^1$, $X\cong D^3\times S^1$ and $Y\cong S^2\times D^2$. When $m=1$ the link is the Borromean rings Bo, and X is a regular neighbourhood of the unknotted embedding of the torus T in S^4. When $m=-1$, the link is 8^3_9, and X is a regular neighbourhood of the unknotted embedding of the Klein bottle Kb in S^4 with normal Euler number 0. In general, X is aspherical, $\pi_X\cong BS(1,m)$ and $\pi_Y\cong \mathbb{Z}/(m-1)\mathbb{Z}$. (Note, however, that the boundary of a regular neighbourhood of the Fox 2-knot with group $BS(1,2)$ gives an embedding of $S^2\times S^1$ with $\pi_X\cong BS(1,2)$ and $\chi(X)=0$, but this embedding is not bi-epic and X is not aspherical.)

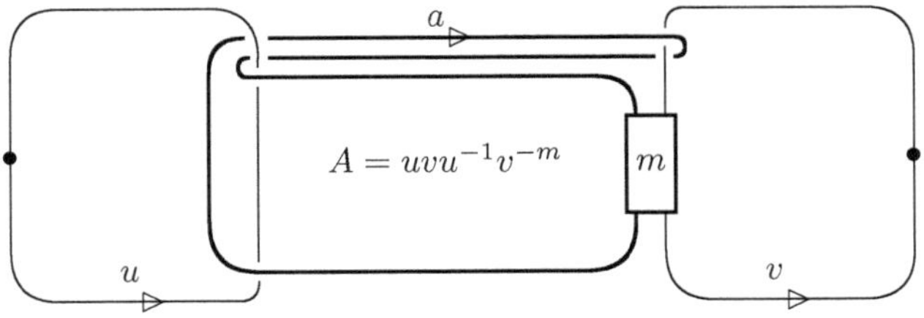

Figure 7.3 $\pi_X\cong BS(1,m)$.

If W aspherical and π_W is elementary amenable (but is not a PD_3-group) then either $W\simeq *$ or $\pi_W\cong\mathbb{Z}$ or $BS(1,m)$ [**DH25**]. There are no other known (finitely presentable) restrained groups G with $c.d.G=2$, and certainly no others which are almost coherent and have infinite abelianization [**FMGK**, Theorem 2.6].

If W is aspherical and π_W is a non-trivial amenable group then $\chi(W)=0$, while if $\pi_W = F(r)$ for some $r>0$ then $\chi(W)\leqslant 0$. In either of these cases $W=X$, by our convention on labelling the regions. If the complementary region Y is aspherical then $\chi(\pi_Y)>0$, and so π_Y is neither amenable nor free.

If X is aspherical and $\pi_X \cong F(r)$ is free then $H_1(X)$ is torsion-free and $H_2(X) = 0$, and so $H_1(Y) = 0$ and $H_2(Y) \cong \mathbb{Z}^r$. This situation is realized by the standard embedding of $\#^r(S^2 \times S^1)$. Are there examples with Y also aspherical?

7.9. Recognizing the Simplest Embeddings

The simplest 3-manifolds to consider in the present context are perhaps the total spaces of S^1-bundles over orientable surfaces. Most of those which embed have canonical "simplest" embeddings. We give some evidence that these may be characterized up to s-concordance by the conditions $\pi_X \cong \pi_1 F$, where F is the base, and π_Y is abelian.

Suppose first that $M \cong T_g \times S^1$. There is a canonical embedding $j_g : M \to S^4$, as the boundary of a regular neighbourhood of the standard smooth embedding $T_g \subset S^3 \subset S^4$. Let X_g and Y_g be the complementary components. Then $X_g \cong T_g \times D^2$ and $Y_g \simeq S^1 \vee \bigvee^{2g} S^2$, and so $\pi_{Y_g} \cong \mathbb{Z}$.

We shall assume henceforth that $g \geqslant 1$, since embeddings of $S^2 \times S^1$ and $S^3 = M(0; (1,1))$ may be considered well understood. Let h be the image of the fibre in π.

LEMMA 7.14. *Let $j : T_g \times S^1 \to S^4$ be an embedding such that $\pi_X \cong \pi_1 T_g$. Then X is s-cobordant rel ∂ to $X_g = T_g \times D^2$.*

PROOF. Since $H^2(X) \cong \mathbb{Z}$ is a direct summand of $H^2(M)$ and is generated by cup products of classes from $H^1(X)$, the image of j_{X*} cannot be a free group. Therefore, the index d of this image in π_X is finite, and so $\chi(\mathrm{Im}(j_{X*})) = d\chi(F)$. Since $\mathrm{Im}(j_{X*})$ is an orientable surface group, it requires at least $2 - d\chi(F) = 2(gd - d + 1)$ generators. On the other hand, π needs just $2g + 1$ generators. Thus if $g > 1$ we must have $d = 1$, and so j_{X*} is onto. This is also clear if $g = 1$, for then $\pi_X \cong H_1(X)$ is a direct summand of $H_1(M)$. In all cases, we may apply Theorem 7.10 to conclude that X is aspherical.

Any homeomorphism from ∂X to ∂X_g which preserves the product structure extends to a homotopy equivalence of pairs $(X, \partial X) \simeq (X_g, \partial X_g)$. Now $L_5(\pi_1 T_g)$ acts trivially on the s-cobordism structure set $S^s_{TOP}(X_g, \partial X_g)$, by Theorem 6.7 and Lemma 6.9 of [**FMGK**]. Since T_g has a metric of constant non-positive curvature the integral Novikov conjecture holds for $\pi_X \cong \pi_1 T_g$, and since X is aspherical it follows that $\sigma_4(X, \partial X)$ is an isomorphism. (See [**KL**], especially sections 9.2, 20.2 and 24.1.) Therefore X and X_g are TOP s-cobordant (rel ∂). □

If $\pi_Y \cong \mathbb{Z}$ then $\Sigma = Y \cup (T_g \times D^2)$ is 1-connected, since π_Y is generated by the image of h, and $\chi(\Sigma) = 2$. Hence Σ is a homotopy 4-sphere, containing a locally flat copy of T_g with exterior Y.

LEMMA 7.15. *If there is a map $f : Y \to Y_g$ which extends a homeomorphism of the boundaries then Y is homeomorphic to Y_g.*

PROOF. Let t be a generator t for π_Y. Then $\mathbb{Z}[\pi_Y] \cong \Lambda = \mathbb{Z}[t, t^{-1}]$. Let $\Pi = \pi_2 Y$. As in Theorem 7.10, $H_q(Y; \Lambda) = H^q(Y; \Lambda) = 0$ for $q > 2$, and the equivariant chain complex for $\widetilde{Y}$ is chain homotopy equivalent to a finite projective Λ-complex

$$Q_* = \Pi \oplus (Z_1 \to Q_1 \to Q_0)$$

of length 2, with $Z_1 \to Q_1 \to Q_0$ a resolution of $\mathbb{Z}$. The alternating sum of the ranks of the modules Q_i is $\chi(Y) = 2g$. Hence $\Pi \cong \Lambda^{2g}$, since projective Λ-modules are free. In particular, this holds also for Y_g.

If $f : Y \to Y_g$ restricts to a homeomorphism of the boundaries then $\pi_1 f$ is an isomorphism. Comparison of the long exact sequences of the pairs shows that f induces an isomorphism $H_4(Y, \partial Y) \cong H_4(Y, \partial Y)$, and so has degree 1. Therefore, $\pi_2 f = H_2(f; \Lambda)$ is onto, by Poincaré-Lefshetz duality. Since $\pi_2 Y$ and $\pi_2 Y_g$ are each free of rank $2g$, it follows that $\pi_2 f$ is an isomorphism, and so f is a homotopy equivalence, by the Whitehead and Hurewicz Theorems.

Thus f is a homotopy equivalence *rel* ∂, by the Homotopy Extension Property, and so it determines an element of the structure set $\mathcal{S}_{TOP}(Y_g, \partial Y_g)$. The group $L_5(\mathbb{Z})$ acts trivially on the structure set, as in Lemma 7.14, and so the normal invariant gives a bjection $\mathcal{S}_{TOP}(Y_g, \partial Y_g) \cong H^2(Y_g, \partial Y_g; \mathbb{F}_2) \cong H_2(Y_g; \mathbb{F}_2)$. Since $H_2(\mathbb{Z}; \mathbb{F}_2) = 0$ the Hurewicz homomorphism maps $\pi_2 Y_g$ onto $H_2(Y_g; \mathbb{F}_2)$. Therefore, there is an $\alpha \in \pi_2 Y_g$ whose image in $H_2(Y_g; \mathbb{F}_2)$ is the Poincaré dual of the normal invariant of f. Let f_α be the composite of the map from Y_g to $Y_g \vee S^4$ which collapses the boundary of a 4-disc in the interior of Y_g with $id_{Y_g} \vee \alpha\eta^2$, where η^2 is the generator of $\pi_4(S^2)$. Then f_α is a self-homotopy equivalence of $(Y_g, \partial Y_g)$ whose normal invariant agrees with that of f [**CH90**]. Therefore, f is homotopic to a homeomorphism $Y \cong Y_g$. □

However, finding such a map f to begin with seems difficult. Can we somehow use the fact that Y and Y_g are subsets of S^4?

Suppose now that W is an s-cobordism *rel* ∂ from X to X_g, and that $Y \cong Y_g$. Since $g \geqslant 1$ the 3-manifold $T_g \times S^1$ is irreducible and sufficiently large. Therefore, $\pi_0(Homeo(T_g \times S^1)) \cong Out(\pi)$ [**Wd68**]. If $g > 1$ then the centre of $\pi_1 T_g$ is trivial, and so $Out(\pi) \cong \begin{pmatrix} Out(\pi_1 T_g) & 0 \\ \mathbb{Z}^{2g} & \mathbb{Z}^\times \end{pmatrix}$. It follows easily that every self homeomorphism of $T_g \times S^1$ extends to a self-homeomorphism of $T_g \times D^2$. Attaching $Y \times [0,1] \cong Y_g \times [0,1]$ to W along $T_g \times S^1 \times [0,1]$ gives an s-concordance from j to j_g.

If $g = 1$ then $X \cong T \times D^2$ and $Out(\pi) \cong GL(3, \mathbb{Z})$. Automorphisms of π are generated by those which may be realized by homeomorphisms of $T \times D^2$ together with those that may be realized by homeomorphisms of Y_1 [**Mo83**]. Thus, if embeddings of T with group $\mathbb{Z}$ are standard so are embeddings of $S^1 \times S^1 \times S^1$ with both complementary components having abelian fundamental groups.

The situation is less clear for bundles over T_g with Euler number ± 1. We may construct embeddings of such manifolds by fibre sum of an embedding of $T_g \times S^1$ with the Hopf bundle $\eta : S^3 \to S^2$. However, it is not clear how the complements change under this operation. There are natural 0-framed links representing such bundle spaces. Since the Whitehead link Wh is an interchangeable 2-component link, $M(1; (1,1)) = M(Wh)$ has an embedding with $X \cong Y \simeq S^1 \vee S^2$ and $\pi_X \cong$

$\pi_Y \cong \mathbb{Z}$. Is this embedding characterized by these conditions? (Once again, it is enough to find a map which restricts to a homeomorphism on boundaries.)

The 3-manifold $\#^\beta(S^2\times S^1)$ is the result of 0-framed surgery on the β-component trivial link βU, and so has embeddings realizing all the possibilities for Euler characteristics allowed by Lemma 2.2. In particular, it has a "standard" embedding $j_{\beta U}$ with complementary regions $X \cong \natural^\beta(D^3 \times S^1)$ and $Y \cong \natural^\beta(S^2 \times D^2)$.

THEOREM 7.16. *Let $M = \#^\beta(S^2 \times S^1)$ and let $j : M \to S^4$ be an embedding such that $\chi(X) = 1 - \beta$ and j_{X*} is an epimorphism. Then j is s-concordant to the standard embedding $j_{\beta U}$.*

PROOF. Since $\pi \cong F(\beta)$ is a free group and $H_2(X) = 0$, the homomorphism from π to $\pi_X/\bigcap_{n\geqslant 1}\gamma_n\pi_X$ is a monomorphism, by Lemma 2.4 and the residual nilpotence of free groups [**Rob**, 6.10.1]. Hence j_{X*} is an isomorphism, so $\pi_Y = 1$, by Lemma 2.5, and j is bi-epic. Therefore $c.d.X \leqslant 2$ and $c.d.Y \leqslant 2$, by Theorem 7.10.

The inclusion of $M = \#^\beta(S^2 \times S^1)$ into $\natural^\beta(S^2 \times D^2) \simeq \vee^\beta S^2$ induces an isomorphism $H^2(\vee^\beta S^2) \cong H^2(M)$, and we see easily that it extends to a map $g : (Y, M) \to (\natural^\beta(S^2\times D^2), \#^\beta(S^2\times S^1))$. Since Y is 1-connected, $c.d.Y \leqslant 2$ and $H^2(g)$ is an isomorphism, g is a homotopy equivalence. Moreover, $w_2(Y) = 0$ and Y has signature 0, since it is a subset of S^4. As in Theorem 7.2, any such map is homotopic *rel* ∂ to a homeomorphism, by 1-connected surgery, and so $Y \cong \natural^\beta(S^2 \times D^2)$.

Since $\chi(X) = 1 - \beta$ it follows easily that X is aspherical. Hence $X \simeq \vee^\beta S^1$, and so (X, M) is s-cobordant *rel* ∂ to $(\natural^\beta(D^3\times S^1), \#^\beta(S^2\times S^1))$ by [**FQ**, Theorem 11.6A]. Every self-homeomorphism of $\#^\beta(S^2\times S^1)$ extends across $\natural^\beta(D^3\times S^1)$, and so j is s-concordant to the standard embedding $j_{\beta U}$. □

We remark that if $M = M_1\#M_2$ is a proper connected sum of 3-manifolds which embed in S^4 and one of the summands has embeddings with differing values of $\chi(X)$, then so does M.

CHAPTER 8

Abelian Embeddings

We begin this chapter with a simple homological argument which severely restricts the abelian possibilities for π_X. Homology 3-spheres have essentially unique abelian embeddings (although they may have other embeddings). This is also known for $S^2 \times S^1$ and $S^3/Q(8)$, by results of Aitchison [**Rub80**] and Lawson [**Law84**], respectively. In Theorems 8.8 and 8.9 we show that if M is an orientable homology handle (i.e., such that $H_1(M) \cong \mathbb{Z}$) then it has an abelian embedding if and only if the commutator subgroup of $\pi_1 M$ is perfect, and then the abelian embedding is essentially unique. (There are homology handles which do not embed in S^4 at all!) The 3-manifolds obtained by 0-framed surgery on 2-component links with unknotted components always have abelian embeddings, and the complementary regions for such embeddings are homotopy equivalent to standard 2-complexes. These shall be our main source of examples. In particular, we shall give an example in which $X \simeq Y \simeq S^1 \vee S^2$, but the pairs (X, M) and (Y, M) are not homotopy equivalent. (See Appendix C and [**Hi25**] for examples of 3-manifolds with several inequivalent abelian embeddings.)

8.1. Constraints on the Invariants

In this section we shall show that manifolds with embeddings for which π_X is abelian are severely constrained.

THEOREM 8.1. *Suppose M has an embedding in S^4 for which one complementary region X has $\chi(X) \leqslant 1$ and $\gamma_2\pi_X = \gamma_3\pi_X$. Then either $\beta \leqslant 4$ or $\beta = 6$. If $\beta = 0$ or 2 then $\pi_X \cong \mathbb{Z}/n\mathbb{Z}$ or $\mathbb{Z} \oplus \mathbb{Z}/n\mathbb{Z}$, respectively, for some $n \geqslant 1$, while if $\beta = 1$, 3, 4 or 6 then $\pi_X \cong \mathbb{Z}^\gamma$, where $\gamma = \lfloor \frac{\beta+1}{2} \rfloor$. If π_X is abelian and $\beta = 1$ or 3 then X is aspherical.*

PROOF. Let $\gamma = \beta_1(X)$ and $A = H_1(X)$. Then $2\gamma \geqslant \beta$ and $A \cong \mathbb{Z}^\gamma \oplus \tau_X$. Since A is abelian, $H_2(A) = A \wedge A \cong \mathbb{Z}^{\binom{\gamma}{2}} \oplus (\tau_X)^\gamma \oplus (\tau_X \wedge \tau_X)$.

If $\gamma_2\pi_X = \gamma_3\pi_X$ then $H_2(A)$ is a quotient of $H_2(\pi_X)$, by the 5-term exact sequence of low degree for π_X as an extension of A. This in turn is a quotient of $H_2(X) \cong \mathbb{Z}^{\beta-\gamma}$, by Hopf's Theorem. Hence $\binom{\gamma}{2} \leqslant \beta - \gamma \leqslant \gamma$, and so $\gamma \leqslant 3$. If $\tau_X \neq 0$ then either $\gamma = \beta = 0$ and $\tau_X \wedge \tau_X = 0$, or $\gamma = 1$, $\beta = 2$ and $\tau_X \wedge \tau_X = 0$. In either case, τ_X is (finite) cyclic. If $\beta \neq 0$ or 2 then $\tau_X = 0$ and either $\gamma = \beta = 1$, or $\gamma = 2$ and $\beta = 3$ or 4, or $\gamma = 3$ and $\beta = 6$. The final assertion follows immediately from Theorem 7.10, since $cd\pi_X = \gamma \leqslant 2$ and $\chi(X)$ must be 0. □

If π_X is abelian, $\gamma = \beta = 0$ and $\tau_M = 0$ then X is contractible. In the remaining cases X cannot be aspherical, since either π_X has non-trivial torsion (if $\beta = 0$), or $H_2(X)$ is too big (if $\beta = 2$ or 4), or $H_3(X)$ is too small (if $\beta = 6$).

If we assume merely that $\gamma_2\pi_X/\gamma_3\pi_X$ is finite (i.e., that the rational lower central series stabilizes after one step) then $\cup_X : \wedge_2 H^1(X;\mathbb{Q}) \to H^2(X;\mathbb{Q})$ is injective [**Su75**], and a similar calculation gives the same restrictions on β.

COROLLARY 8.1.1. *Let M be a 3-manifold with an abelian embedding in S^4. Then*

(1) *if $\beta = 3$ then there is a $\mathbb{Z}$-homology isomorphism $h : M \to T^3$;*
(2) *if $\beta = 4$ or 6 then $\mu_M \neq 0$.*

PROOF. The homomorphism induced by cup product from $\wedge_2 H^1(X)$ to $H^2(X)$ is injective, since π_X is abelian [**Su75**].

If $\beta = 3$ then $\pi_X \cong \mathbb{Z}^2$ and X is aspherical, by Theorem 8.1. Hence $X \simeq T^2$, and so $H^1(X)$ has a basis a, b such that $a \cup b$ generates a free summand of $H^2(M)$. We may then choose $c \in H^1(M)$ so that $(a \cup b \cup c)[M] = 1$. If we identify $H^1(M)$ with the homotopy set $[M; S^1]$ we see that a, b, c together determine a map h from M to T^3 which is an isomorphism on $H^1(M)$ and on $H^3(M)$. This is clearly a $\mathbb{Z}$-homology isomorphism.

If $\beta \geqslant 3$ then $\beta_1(X) \geqslant 2$, and so there are $a, b \in H^1(X)$ with $a \cup b \neq 0$ in $H^2(M)$. Hence there is a $c \in H^1(M)$ such that $a \cup b \cup c \neq 0$, by Poincaré duality for M. □

In particular, $\#^\beta(S^2 \times S^1)$ has an abelian embedding if and only if $\beta \leqslant 2$ (in which case $j_{\beta U}$ is abelian).

Embeddings with π_X abelian and $\beta \leqslant 4$ realizing these possibilities may be easily found. (If $\pi_X \neq 1$ then 2-knot surgery gives further examples with π_X non-abelian and $\gamma_2\pi_X = \gamma_3\pi_X$.) The simplest examples are for $\beta = 0, 1$ or 3, with $M \cong S^3$, $M = S^2 \times S^1$ or $S^1 \times S^1 \times S^1$ the boundary of a regular neighbourhood of a point or of the standard unknotted embedding of S^2 or T in S^4, respectively. Other examples may be given in terms of representative links. When $\beta = 0$ the $(2, 2n)$ torus link gives examples with $X \cong Y$ and $\pi_X \cong \mathbb{Z}/n\mathbb{Z}$. When $\beta = 1$ we may use any knot which bounds a slice disc $D \subset D^4$ such that $\pi_1(D^4 \setminus D) \cong \mathbb{Z}$, such as the unknot or the Kinoshita-Terasaka knot 11_{n42}. (All such knots have Alexander polynomial 1. Conversely every Alexander polynomial 1 knot bounds a TOP locally flat slice disc with group $\mathbb{Z}$, by a striking result of Freedman.) The links 8^3_5 and 8^3_6 give further simple examples. (These each have a trivial 2-component sublink and an unknotted third component which represents a meridian of the first component or the product of meridians of the first two components, respectively.) When $\beta = 2$ any 2-component link with unknotted components and linking number 0, such as the trivial 2-component link or Wh, gives examples with $\pi_X \cong \mathbb{Z}$. We may construct examples realizing $\mathbb{Z} \oplus \mathbb{Z}/n\mathbb{Z}$ from the 4-component link obtained from the Borromean rings Bo by replacing one component by its $(2, 2n)$ cable. When $\beta = 3$ we may use the links Bo, 9^3_9 or 9^3_{18}. (These each have a trivial 2-component sublink and an unknotted third component which represents the commutator of the meridians of the first two components. However, neither of the latter two links is Brunnian.)

EXAMPLE 8.2. $\beta = 4$.
Let L be the 4-component link obtained from Bo by adjoining a parallel to the third component, and let M be the 3-manifold M obtained by 0-framed surgery on L. Then the meridians of L represent a basis $\{e_i\}$ for $H_1(M) \cong \mathbb{Z}^4$, and $\mu_M = e_1 \wedge e_2 \wedge e_3 + e_1 \wedge e_2 \wedge e_4$. This link has two essentially different partitions as the union of trivial 2-component links, and ambient surgery gives two essentially different embeddings of M. If the sublinks are $\{L_1, L_2\}$ and $\{L_3, L_4\}$ then the complementary components have fundamental groups $\mathbb{Z}^2$ and $F(2)$. Otherwise, the complementary components are homeomorphic and have fundamental group $\mathbb{Z}^2$.

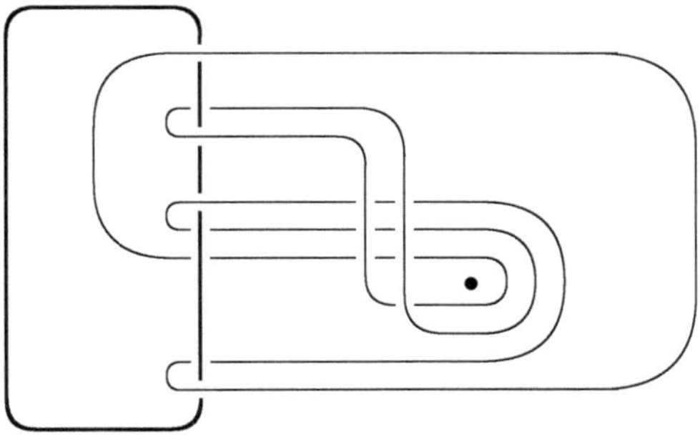

Figure 8.1 A version of Wh with branching axis through $\bullet$.

EXAMPLE 8.3. $\beta = 6$.
The link diagram in Figure 8.1 is a projection of Wh. The pre-images of this link in the 2- and 3-fold branched cyclic covers of S^3, branched over the trivial knot represented in cross-section by $\bullet$, are 4- and 6-component bipartedly trivial links in which each component represents the commutator of meridians of the adjacent components (as in Bo). The corresponding 3-manifolds have $\beta = 4$ and 6, respectively, and the associated embeddings are abelian. These 3-manifolds are branched cyclic covers of $M(Wh)$, but we are not aware of any other features which are of particular interest. (Theorem 2.8 also provides examples for the case $\beta = 6$.)

In all of the above examples except for one, π_Y is also abelian. Note that Theorem 8.1 does *not* apply to π_Y, as it uses the hypothesis $\beta_1(X) \geqslant \frac{1}{2}\beta$.

LEMMA 8.4. *If π_X is abelian of rank at most* 1 *then X is homotopy equivalent to a finite* 2*-complex. If π_X is cyclic then $X \simeq S^1$, $S^1 \vee S^2$ or $P_\ell = S^1 \vee_\ell e^2$, for some $\ell \neq 0$.*

PROOF. The first assertion follows from the facts that $c.d.X \leqslant 2$, by Theorem 7.10, and that the $\mathcal{D}(2)$ property holds for cyclic groups (see [**John**, page 235]) and for the groups $\mathbb{Z} \oplus \mathbb{Z}/\ell\mathbb{Z}$ [**Ed06**]. If $\pi_X \cong \mathbb{Z}/\ell\mathbb{Z}$ is cyclic then $X \simeq S^1$ or $S^1 \vee S^2$, if $\ell = 0$, or $P_\ell = S^1 \vee_\ell e^2$, if $\ell \neq 0$ [**DS73**]. □

We shall show later that a similar result holds when $\pi_X \cong \mathbb{Z}^2$.

Ten of the thirteen 3-manifolds with elementary amenable fundamental groups and which embed in S^4 have abelian embeddings. In at least four cases (S^3, $S^3/Q(8)$, S^3/I^* and $S^2 \times S^1$) the abelian embedding is essentially unique. The flat 3-manifold $M(-2;(1,0))$ and the $\mathbb{N}il^3$-manifold $M(-2;(1,4))$ bound regular neighbourhoods of embeddings of the Klein bottle Kb in S^4, but have no abelian embeddings. The status of the $\mathbb{S}ol^3$-manifold $M_{2,4}$ is not yet known.

When $\beta \leqslant 1$ we must have $\chi(X) = 1 - \beta$. If L is a 2-component slice link with unknotted components (such as the trivial 2-component link, or the Milnor boundary link) and $M = M(L)$ then $\beta = 2$ and M has an abelian embedding (with $\chi(X) = \chi(Y) = 1$), and also an embedding with $\chi(X) = 1 - \beta = -1$ and $\pi_Y = 1$. However, it shall follow from the next lemma that if $\beta > 2$ then M cannot have both an abelian embedding and also one with $\chi(X) = 1 - \beta$.

LEMMA 8.5. *Let $j : M \to S^4$ be an embedding j such that $H_1(Y) = 0$, and let $S \subset \Lambda_\beta = \mathbb{Z}[\pi/\pi']$ be the multiplicative system consisting of all elements s with augmentation $\varepsilon(s) = 1$. If the augmentation homomorphism $\varepsilon : \Lambda_{\beta S} \to \mathbb{Z}$ factors through an integral domain $R \neq \mathbb{Z}$ then $H_1(M;R)$ has rank $\beta - 1$ as an R-module.*

PROOF. Let $*$ be a basepoint for M and $A(\pi) = H_1(M,*;\Lambda_\beta)$ be the Alexander module of π. (See [**AIL**, Chapter 4].) Since $H_2(X) = 0$, the inclusion of representatives for a basis of $H_1(X) \cong \mathbb{Z}^\beta$ induces isomorphisms $F(\beta)/\gamma_n F(\beta) \cong \pi/\gamma_n\pi$, for all $n \geqslant 1$, by Lemma 2.4. Hence $A(\pi)_S \cong (\Lambda_{\beta S})^\beta$, by [**AIL**, Lemma 4.9]. Since ε factors through R, the exact sequence of the pair $(M,*)$ with coefficients R gives an exact sequence

$$0 \to H_1(M;R) \to R \otimes_{\Lambda_\beta} A(\pi) \cong R^\beta \to R \to R \otimes_{\Lambda_\beta} \mathbb{Z} = \mathbb{Z} \to 0,$$

from which the lemma follows. (Note that the hypotheses on R imply that $\mathbb{Z}$ is an R-torsion module.) □

8.2. Homology Spheres

We begin this section by showing that Freedman's Embedding Theorem for homology spheres is an easy application of the heavy machinery of surgery.

THEOREM 8.6 (Freedman). [**FQ**, §11.4] *Let Σ be a homology sphere. Then Σ has an unique abelian embedding in S^4.*

PROOF. Let $C = c\Sigma$ be the cone on Σ. Then $H_4(C,\Sigma) \cong H_3(\Sigma) \cong \mathbb{Z}$. Cap product with a generator $[C,\Sigma]$ of this group defines isomorphisms from $H^i(C)$ to $H_{4-i}(C,\Sigma)$, for all i, since C is contractible and Σ is a homology sphere. Since $\pi_1(X) = 1$ this is enough to ensure that (C,Σ) is a PD_4-pair. The only obstruction to reducing the stable spherical fibration of this PD_4-pair to a TOP bundle lies in $H^3(C,\Sigma;\mathbb{F}_2) \cong H_1(C,\mathbb{F}_2) = 0$. Hence we may apply surgery to see that $(C,\Sigma) \simeq (W,\Sigma)$, where W is a contractible 4-manifold with boundary Σ. Doubling W along its boundary gives a homotopy 4-sphere, which is homeomorphic to S^4 by the validity of the Poincaré Conjecture, and so Σ embeds in S^4.

This embedding is clearly abelian, since $\pi_X = \pi_Y = 1$. More generally, if j is any abelian embedding of Σ then the complementary regions X and Y are contractible. Homeomorphisms of boundaries $\partial X \cong \partial W$ and $\partial Y \cong \partial W$ extend to homotopy equivalences of pairs $(X, \Sigma) \simeq (W, \Sigma)$ and $(Y, \Sigma) \simeq (W, \Sigma)$. The surgery exact sequence then shows that (X, Σ) and (Y, Σ) are homeomorphic to (W, Σ), and that the abelian embedding is unique. □

When $M = S^3$, the result goes back to the Generalized Schoenflies Theorem, which does not use surgery. In this special case the embedding is essentially unique!

It is not clear whether homology spheres other than S^3 must have embeddings with one or both of π_X and π_Y non-trivial. Figure 2.2 gives an example with $\pi_X \cong \pi_Y \cong I^*$, the binary icosahedral group. In this case the homology sphere is the result of surgery on a complicated 4-component bipartedly trivial link, and probably has no simpler description. The Poincaré homology sphere S^3/I^* is not the result of 0-framed surgery on any bipartedly slice link, since it does not embed smoothly.

We mention here several results from [**Liv05**] on smooth embeddings of homology spheres. There is a homology sphere which has at least two smooth embeddings; one with X contractible and one with $\pi_X \neq 1$. There is a superperfect group which is not the fundamental group of a homology 4-ball. There is a contractible codimension-0 submanifold $W \subset S^4$ such that $def(\pi_W) < -n$, for every $n \geqslant 0$. Moreover, DW is diffeomorphic to S^4.

If $S^4 = DW$ is the double of a 4-manifold W with connected boundary $\partial W = M$ then W is 1-connected, by the Van Kampen Theorem, and so contractible. Thus, M must be a homology sphere. Conversely, if M is a homology sphere and X and Y are contractible then there is a homeomorphism $h : X \cong Y$ such that $j_Y = h \circ j_X$, and so $S^4 \cong DX$. In general one may construct examples with $Y \cong X$ and β any positive even number.

EXAMPLE 8.7. *β even and $Y \cong X$.*
Let $p : S^3 \to S^3$ be a $2d$-fold branched cyclic covering, with branch set the first component of the Whitehead link. The preimage of the second component is a $2d$-component link L which is the union of two trivial d-component links, and any generator of the covering group carries one sublink onto the other. (When $d = 1$ the preimage of the whole Whitehead link is the 3-component link 8^3_9, which is the union of a copy of the $(2, 4)$-torus link 4^2_1 and an unknot.) Thus, we obtain an example with $\beta = 2d$ and $X \cong Y$.

In such cases S^4 is a twisted double $X \cup_\psi X$, for some self-homeomorphism of $M = \partial X$. See [**Ya97**] for other representations of S^4 as a twisted double.

If $j : M \to S^4$ is an embedding such that one of the complementary regions is 1-connected then $H_2(X) = 0$, and so $\pi/\gamma_n\pi \cong F(\beta)/\gamma_n F(\beta)$, for all n, by Lemma 2.4. The latter condition holds if $M = M(L)$ for some slice link, and clearly the embedding deriving from a partition of such a link L as $L_+ \cup L_-$ with L_- empty has $\pi_Y = 1$.

8.3. Homology Handles

If M is a homology handle then there is a $\mathbb{Z}$-homology isomorphism $f : M \to S^2 \times S^1$, by Lemma 8.9, and π'/π'' is a finitely generated torsion module over $\mathbb{Z}[\pi/\pi'] \cong \Lambda = \mathbb{Z}[t, t^{-1}]$.

THEOREM 8.8. *Let M be an orientable homology handle. If M embeds in S^4 then the Blanchfield pairing on $\pi'/\pi'' = H_1(M; \mathbb{Z}[\pi/\pi'])$ is neutral. There is an abelian embedding $j : M \to S^4$ if and only if π' is perfect, and then $X \simeq S^1$ and $Y \simeq S^2$.*

PROOF. The first assertion follows on applying equivariant Poincaré duality to the infinite cyclic cover of the pair (X, M). (See the proof of [**AIL**, Theorem 2.4].)

If j is abelian then $\pi_X \cong \mathbb{Z}$ and $\pi_Y = 1$, while $H_2(X) = 0$ and $H_2(Y) \cong \mathbb{Z}$. Since $c.d.X \leqslant 2$ and $\pi_X \cong \mathbb{Z}$, it follows that $\pi_2 X = H_2(X; \mathbb{Z}[\pi_X])$ is a free $\mathbb{Z}[\pi_X]$-module of rank $\chi(X) = 0$. Hence $\pi_2 X = 0$, and so maps $f : S^1 \to X$ and $g : S^2 \to Y$ representing generators for π_X and $\pi_2 Y$ are homotopy equivalences. Since $H_2(X, M; \mathbb{Z}[\pi_X]) \cong \overline{H^2(X; \mathbb{Z}[\pi_X])} = 0$, by equivariant Poincaré duality, $\pi'/\pi'' = H_1(M; \mathbb{Z}[\pi_X]) = 0$, by the homology exact sequence for the infinite cyclic cover of the pair (X, M). Hence π' is perfect.

Suppose, conversely, that π' is perfect. Then M embeds in S^4, by Corollary 6.12.1, and an examination of the proof shows that the embedding constructed in Theorem 6.12 is abelian. □

If a Seifert manifold M is a homology handle other than $S^2 \times S^1$ then there is no Λ-homology isomorphism from M to $S^2 \times S^1$, by Corollary 3.3.1. It then follows from Theorem 8.8 that if M has an abelian embedding then $M \cong S^2 \times S^1$.

COROLLARY 8.8.1. *If K is a knot then j_K is an abelian embedding if and only if K has Alexander polynomial 1.* □

If K is a knot such that $M = M(K)$ embeds and W is the trace of the surgery from S^3 to M then $\Delta = X \cup_M W$ is a 1-connected homology 4-ball with boundary S^3, and so is homeomorphic to D^4. Hence K is TOP slice. However, if K is a slice knot with non-trivial Alexander polynomial then no embedding is abelian. There are obstructions beyond neutrality of the Blanchfield pairing to slicing a knot, which probably also obstruct embeddings of homology handles.

THEOREM 8.9. *Let M be an orientable homology handle. Then M has at most one abelian embedding, up to equivalence.*

PROOF. Assume that j and j_1 are abelian embeddings of M. There is a homotopy equivalence of pairs $f : (X_1, M) \simeq (X, M)$ which extends id_M, by Lemma 7.9. The normal invariant map ν from $\mathcal{S}_{TOP}(X, \partial X)$ to $\mathcal{N}(X, \partial X)$ is constant, since $H_2(X; \mathbb{F}_2) = 0$, and the surgery obstruction group $L_5(\mathbb{Z})$ acts trivially on the structure set $\mathcal{S}_{TOP}(X, \partial X)$. (This follows from the Wall-Shaneson Theorem and the existence of the E_8-manifold [**FMGK**, Theorem 6.7].) Hence f is homotopic *rel* M to a homeomorphism F.

The other complementary components Y and Y_1 are homotopy equivalent to S^2. Since they are codimension-0 submanifolds of S^4 the intersection pairings on $H_2(Y)$ and $H_2(Y_1)$ are trivial, and $w_2(Y_1) = w_2(Y) = 0$. It then follows from [**Bo86**, Proposition 0.5] that there is a homeomorphism $G : Y_1 \cong Y$ such that $G \circ j_{Y_1} = j_Y$. The map $h = F \cup G$ is a homeomorphism of S^4 such that $hj_1 = j_2$. □

An alternative approach to identifying Y would be to use the fact that $w_2(Y) = 0$ and the pinch construction based on the generator of $\pi_4(S^2)$ to show that there is a self-homotopy equivalence *rel* ∂ of the pair (Y, M) with non-trivial normal invariant, as in [**Law84**, §3].

EXAMPLE 8.10. *$M = M(11_{n42})$ has an essentially unique abelian embedding, although $M = M(K)$ for infinitely many distinct knots K.*

The knot 11_{n42} is the Kinoshita-Terasaka knot, which is the simplest non-trivial knot with Alexander polynomial 1. This bounds a smoothly embedded disc D in D^4, such that $\pi_1(D^4 \setminus D) \cong \mathbb{Z}$, obtained by desingularizing a ribbon disc. (See [**AIL**, Figure 1.4].) Hence M has a smooth abelian embedding. Since 11_{n42} has unkotting number 1, it has an annulus presentation, and so there are infinitely many knots K_n such that $M(K_n) \cong M$ [**AJOT13**]. These knots must all have Alexander polynomial 1, and so each determines an abelian embedding. Are all of these embeddings smoothable, and if so, are they smoothly equivalent?

THEOREM 8.11. *Let K be a 1-knot. Then $M = M(K)$ has a bi-epic embedding if and only if K is homotopically ribbon.*

PROOF. If K is homotopically ribbon then the embedding corresponding to the slice disc demonstrating this property is clearly bi-epic.

Suppose that M has a bi-epic embedding. Let W be the trace of 0-framed surgery on K. Then W is 1-connected, $\chi(W) = 1$ and $\partial W = S^3 \amalg M$. Let $P = X \cup_M W$. Then P is 1-connected, since $\pi(j_X)$ is an epimorphism, $\chi(P) = 1$, and $\partial P = S^3$, and so $P \cong D^4$. Clearly K is homotopically ribbon in P. □

If K is a fibred homotopically ribbon knot then the monodromy for the fibration extends over a handlebody [**CG83**]. Hence M bounds a mapping torus X such that the inclusion $M \subset X$ induces an epimorphism from π to $\pi_1 X$ and an isomorphism on the abelianizations. Let Y be the 4-manifold obtained by adjoining a 2-handle to D^4 along K. Then $\Sigma = X \cup_M Y$ is a homotopy 4-sphere, and the inclusion of M into Σ is bi-epic.

For example, if k is a fibred 1-knot with exterior $E(k)$ and genus g, then $K = k\#{-k}$ is a fibred ribbon knot, and $M(K)$ bounds a thickening X of $E(k) \subset S^3 \subset S^4$, which fibres over S^1, with fibre $\natural^g(S^1 \times D^2)$.

In the next theorem we do not assume that M is a homology handle.

THEOREM 8.12. *Let M be a 3-manifold and $j : M \to S^4$ an embedding. If X fibres over S^1 and j_{X*} is an epimorphism then M is a mapping torus, the* *projection* $p : M \to S^1$ *extends to a map from* X *to* S^1, $\chi(X) = 0$, X *is aspherical and* $\pi_X \cong F(r) \rtimes \mathbb{Z}$ *for some* $r \geqslant 0$. *Conversely, if these conditions hold and the integral Novikov conjecture holds for* π_X *then there is an embedding* $\widehat{j} : M \to S^4$ *which is s-concordant to* j *and such that* $\widehat{X}$ *fibres over* S^1.

PROOF. If X fibres over S^1, with fibre F, then $M = \partial X$ is the mapping torus of a self-homeomorphism of ∂F and the projection $p : M \to S^1$ extends to a map from X to S^1. Moreover, $\chi(X) = 0$ and π_X is an extension of $\mathbb{Z}$ by the finitely presentable normal subgroup $\pi_1 F$. Hence $\beta_1^{(2)}(\pi_X) = 0$, by [**Lück**, Theorem 7.2.6]. If j_{X*} is surjective then $c.d.X \leqslant 2$, and conversely, by Theorem 7.10. Hence X is aspherical, by Lemma 7.11. Conversely, if X is aspherical and $c.d.X \leqslant 2$ then $H_1(X, M : \mathbb{Z}[\pi_X]) \cong H^3(X; \mathbb{Z}[\pi_X]) = 0$, by Poincaré duality with coefficients $\mathbb{Z}[\pi_X]$. Hence the preimage of M in $\widetilde{X}$ is connected, and so j_{X*} is surjective, Hence $\pi_1 F$ is free, by [**Bie**, Corollary 6.6], and so $\pi_X \cong F(r) \rtimes \mathbb{Z}$ for some $r \geqslant 0$.

Suppose now that the conditions in the second sentence of the enunciation hold. Let X^∞ be the covering space associated to the subgroup $F(r)$, and let j_{X^∞} be the inclusion of $M^\infty = \partial X^\infty$ into X^∞. Let τ be a generator of the covering group $\mathbb{Z}$. Since X is aspherical and $\pi_1 X^\infty \cong F(r)$, there is a homotopy equivalence $h : X^\infty \to N = \natural^r(S^1 \times D^2)$. Then there is a self-homeomorphism t_N of N such that $t_N h \sim h\tau$. Let $\theta : \partial N \to N$ be the inclusion, and let $\widehat{X} = M(t_N)$ be the mapping torus of t_N. Then there is a homotopy equivalence $\alpha : M^\infty \to \partial N$ such that $\theta\alpha \sim h j_{X^\infty}$, by a result of Stallings and Zieschang. (See [**GK92**, Theorem 2].) We may modify h on a collar neighbourhood of ∂X^∞ so that $h|_{\partial X^\infty} = \alpha$. Hence h determines a homotopy equivalence of pairs $(X, M) \simeq (\widehat{X}, \partial\widehat{X})$. Since M and $\partial\widehat{X}$ are orientable (Haken) manifolds we may further arrange that $h|_M : M \to \partial\widehat{X}$ is a homeomorphism. The group $L_5(\mathbb{Z}[\pi_X])$ acts trivially on the s-cobordism structure set $S^s_{TOP}(\widehat{X}, \partial\widehat{X})$, by [**FMGK**, Theorem 6.7], and $\sigma_4(X, M)$ is an isomorphism (as in Lemma 7.14), since X is aspherical and the integral Novikov conjecture holds for π_X. Hence X and $\widehat{X}$ are s-cobordant *rel* ∂, by hypothesis. The union $\Sigma = \widehat{X} \cup_M Y$ is an homotopy 4-sphere, and so is homeomorphic to S^4. The composite $\widehat{j} : M \subset \widehat{X} \subset \Sigma \cong S^4$ is clearly s-concordant to j and $X(\widehat{j}) = \widehat{X}$ fibres over S^1. □

In particular, if $\beta = 1$ then $\chi(X) = 0$ and M is a rational homology handle. In this case $\widehat{j}$ is equivalent to j, since the s-cobordism theorem holds over $\mathbb{Z}$.

8.4. $\pi/\pi' \cong \mathbb{Z}^2$

When $\pi/\pi' \cong \mathbb{Z}^2$ there is again a simple necessary condition for M to have an abelian embedding.

THEOREM 8.13. *Let* M *be a 3-manifold with fundamental group* π *such that* $\pi/\pi' \cong \mathbb{Z}^2$. *If* $j : M \to S^4$ *is an abelian embedding then* $X \simeq Y \simeq S^1 \vee S^2$, *and* $H_1(M; \mathbb{Z}[\pi_X])$ *and* $H_1(M; \mathbb{Z}[\pi_Y])$ *are cyclic* $\mathbb{Z}[\pi_X]$- *and* $\mathbb{Z}[\pi_Y]$-*modules (respectively), of projective dimension* $\leqslant 1$.

PROOF. Since j is abelian, $\pi_X \cong \pi_Y \cong \mathbb{Z}$ and $\chi(X) = \chi(Y) = 1$. Moreover, since j_{X*} and j_{Y*} are epimorphisms, $c.d.X \leqslant 2$ and $c.d.Y \leqslant 2$, by Theorem 7.10. Hence $X \simeq Y \simeq S^1 \vee S^2$, by Corollary 7.7.1.

We again consider the homology exact sequences of the infinite cyclic covers of the pairs (X, M) and (Y, M), in conjunction with equivariant Poincaré duality. Since $H_i(X; \mathbb{Z}[\pi_X]) = 0$ for $i \neq 0$ or 2 and $H_2(X; \mathbb{Z}[\pi_X]) \cong \mathbb{Z}[\pi_X]$, we have $H_2(X, M; \mathbb{Z}[\pi_X]) \cong \overline{H^2(X; \mathbb{Z}[\pi_X])} \cong \mathbb{Z}[\pi_X]$ also, while $H_3(X, M; \mathbb{Z}[\pi_X]) \cong \overline{H^1(X; \mathbb{Z}[\pi_X])} \cong \mathbb{Z}$. Hence there is an exact sequence

$$0 \to \mathbb{Z} \to H_2(M; \mathbb{Z}[\pi_X]) \to \mathbb{Z}[\pi_X] \to \mathbb{Z}[\pi_X] \to H_1(M; \mathbb{Z}[\pi_X]) \to 0.$$

Therefore, either $H_1(M; \mathbb{Z}[\pi_X]) \cong \mathbb{Z}[\pi_X]$ or $H_1(M; \mathbb{Z}[\pi_X])$ is a cyclic torsion module with a short free resolution. In either case $H_1(M; \mathbb{Z}[\pi_X])$ is a cyclic module of projective dimension $\leqslant 1$.

A similar argument applies for the pair (Y, M). □

To use Theorem 8.13 to show that some M has no abelian embedding we must consider all possible bases for $Hom(\pi, \mathbb{Z})$ or, equivalently, for π/π'.

EXAMPLE 8.14. *Let L be the link obtained from the Whitehead link $Wh = 5_1^2$ by tying $3_1\#-3_1$ in one component. Then no embedding of $M = M(L)$ is abelian.* The link group πL has the presentation

$$\langle a, b, c, r, s, t, u, v, w \mid as^{-1}vsa^{-1} = w = brb^{-1},\ cac^{-1} = b,\ rcr^{-1} = a,\ wcw^{-1} = b,$$
$$rvr^{-1} = tut^{-1},\ sts^{-1} = u,\ usu^{-1} = t,\ vsv^{-1} = r\rangle,$$

and $\pi = \pi_1 M \cong \pi L/\langle\langle \lambda_a, \lambda_r \rangle\rangle$, where $\lambda_a = c^{-1}wr^{-1}a$ and $\lambda_r = vu^{-1}s^{-1}t^{-1}rsa^{-1}b$ are the longitudes of L. Let $b = \beta a$, $c = \gamma a$ and $t = r\tau$. Then $w = \gamma r$ in π, and so π has the presentation

$$\langle a, \beta, \gamma, r, s, \tau, v \mid [r, a] = \gamma^{-1}\beta r\beta^{-1}r^{-1} = r\gamma^{-1}r^{-1},\ \gamma a\gamma^{-1}a^{-1} = \beta,\ sr\tau s = r\tau sr\tau,$$
$$as^{-1}vsa^{-1} = \gamma r,\ vs = rv,\ v = \tau sr\tau s^{-1}\tau^{-1} = \beta^{-1}s^{-1}\tau s^2 r\tau s^{-1}\rangle.$$

Now let $s = \sigma r$ and $v = \xi r$. Then π/π'' has the metabelian presentation

$$\langle a, \beta, \gamma, r, \sigma, \tau, \xi \mid [r, a] = \gamma^{-1}\beta.r\beta^{-1}r^{-1} = r\gamma^{-1}r^{-1},\ \gamma.a\gamma^{-1}a^{-1} = \beta,$$
$$r^{-1}\sigma r.r\tau\sigma r^{-1} = \tau\sigma.r^2\tau r^{-2},\ ar^{-1}\sigma^{-1}\xi ra^{-1}.a\sigma a^{-1} = \gamma.r\gamma^{-1}r^{-1},\ \xi = r\xi\sigma^{-1}r^{-1},$$
$$\xi = \tau\sigma.r^2\tau r^{-2}.r\sigma^{-1}\tau^{-1}r^{-1} = \beta^{-1}.r^{-1}\tau r.\sigma.r^2\tau r^{-2}.r\sigma^{-1}r^{-1},\ [[,],[,]] = 1\rangle,$$

in which $\beta, \gamma, \sigma, \tau$ and ξ represent elements of π', which is the normal closure of the images of these generators. The first relation expresses the commutator $[r, a]$ as a product of conjugates of these generators. Using the third relation to eliminate β, we see that π'/π'' is generated as a module over $\mathbb{Z}[\pi/\pi'] = \mathbb{Z}[a^{\pm}, r^{\pm}]$ by the images of γ, σ, τ and ξ, with the relations

$$(1 - r)[\gamma] = 0,$$
$$(r^2 - r + 1)[\sigma] = r(r^2 - r + 1)[\tau] = 0,$$
$$[\xi] = (1 - r)[\sigma],$$

and

$$2[\sigma] + 2[\tau] = (a - 1)[\gamma].$$

If we extend coefficients to the rationals to simplify the analysis, we see that $P = H_1(M;\mathbb{Q}[\pi/\pi']) = \mathbb{Q}\otimes\pi'/\pi''$ is generated by $[\gamma]$ and $[\tau]$, with the relations

$$(1-r)[\gamma] = (r^2-r+1)[\tau] = 0.$$

Let $\{x,y\}$ be a basis for π/π'. Then $x = a^m r^n$ and $y = a^p r^q$, where $|mq-np| = 1$. Let $\{x^*,y^*\}$ be the Kronecker dual basis for $Hom(\pi,\mathbb{Z})$, and let M_x and M_y be the infinite cyclic covering spaces corresponding to $\mathrm{Ker}(x^*)$ and $\mathrm{Ker}(y^*)$, respectively. Then $H_1(M_x;\mathbb{Q}) \cong (P/(y-1)P\oplus\langle y\rangle)/(x.y = y+[x,y])$. If $H_1(M_x;\mathbb{Q})$ is cyclic as a module over $\mathbb{Q}[x,x^{-1}]$, then so is the submodule

$$P/(y-1)P \cong \mathbb{Q}[\pi/\pi']/(r^2-r+1,y-1)\oplus\mathbb{Q}[\pi/\pi']/(r-1,y-1).$$

On substituting $y = a^p r^q$ we find that this is so if and only if $p = 0$ and $q = \pm 1$. But then $x = a^{\pm 1}$, and a similar calculation shows that $H_1(M_y;\mathbb{Q})$ is not cyclic as a $\mathbb{Q}[y,y^{-1}]$-module. Thus, no basis for π/π' satisfies the criterion of Theorem 8.13, and M has no abelian embedding.

We shall assume henceforth that $M = M(L)$, where L is a 2-component link with components slice knots and linking number $\ell = 0$. Let x and y be the images of the meridians of L in π, and let D_x and D_y be slice discs for the components of L, embedded on opposite sides of the equator $S^3 \subset S^4$. Then the complementary regions for the embedding j_L determined by L are $X_L = (D^4\setminus N(D_x))\cup D_y\times D^2$ and $Y_L = (D^4\setminus N(D_y))\cup D_x\times D^2$. The kernels of the natural homomorphisms from π to π_{X_L} and π_{Y_L} are the normal closures of y and x, respectively. If one of the components of L is unknotted then the corresponding complementary region is a handlebody of the form $S^1\times D^3\cup h^2$. Inverting the handle structure gives a handlebody structure $M\times[0,1]\cup h^2\cup h^3\cup h^4$.

If the components of L are unknotted, or more generally if they have Alexander polynomial 1, then they have slice discs whose complements have fundamental group $\mathbb{Z}$. The embedding j_L is then abelian, with $\pi_X \cong \pi_Y \cong \mathbb{Z}$.

The following example suggests that, beyond this observation, there is little reason to expect simple analogues of Corollary 8.8.1 for embeddings j_L to be abelian when $\beta > 1$. (When $\beta = 1$ and L is a knot, having Alexander polynomial 1 is also a necessary condition for j_L to be abelian, by Corollary 8.8.1.) The link L in Figure 8.2 has components the unknot U and the reef knot $3_1\#-3_1$, and linking number 0. The projection has been drawn to display the ribbon disc for the reef knot. (There are similar examples with components $U, 6_1$.)

The group of the complement of the ribbon disc in Figure 8.2 has the presentation $\langle a,b,c \mid bab^{-1} = c,\ cac^{-1} = b\rangle$ [**AIL**, Theorem 1.15]. The longitude of the unknotted component U is bc^{-1}. Adding a 2-handle to the ribbon disc complement along U gives $\langle a,b,c \mid bab^{-1} = b = c\rangle$, and so this region has fundamental group $\mathbb{Z}$. Since U bounds an unknotted disc in the 4-ball, the embedding j_L is abelian. (The manifold $M(L)$ does not appear to be of independent interest.)

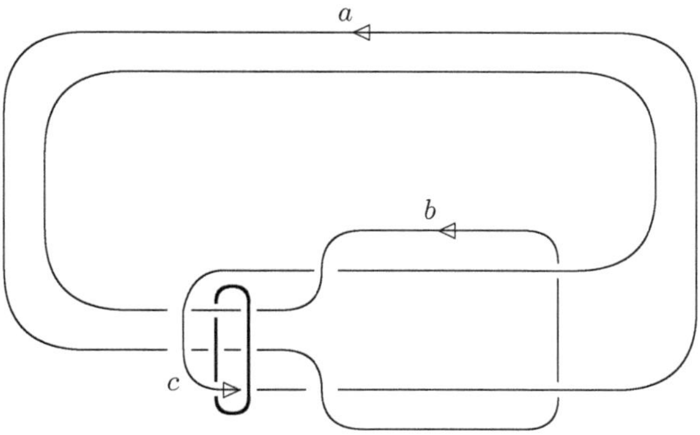

Figure 8.2 $L = (3_1\# - 3_1, U)$.

If a 2-component link L is interchangeable there is a self-homeomorphism of $M(L)$ which swaps the meridians. Hence X_L is homeomorphic to Y_L, and S^4 is a twisted double.

To find examples where the complementary regions are *not* homeomorphic we should start with a link L which is not interchangeable. The simplest condition that ensures that a link with unknotted components is not interchangeable is asymmetry of the Alexander polynomial, and the smallest such link with linking number 0 is 8^2_{13}. Since $\pi = \pi_1 M$ is a quotient of πL, there remains something to be checked.

EXAMPLE 8.15. *The complementary regions of the embedding of* $M = M(8^2_{13})$ *determined by the link* $L = 8^2_{13}$ *are not homeomorphic (although they are homotopy equivalent).*

The group $\pi L = \pi 8^2_{13}$ has the presentation

$$\langle s,t,u,v,w,x,y,z \mid yv = wy,\ zx = wz,\ ty = zt,\ uy = zu,\ sv = us,\ vs = xv,$$

$$wu = tw,\ xs = tx\rangle$$

and the longitudes are $u^{-1}t$ and $x^2z^{-1}ys^{-1}w^{-1}xv^{-1}$. Hence $\pi = \pi_1 M$ has the presentation

$$\langle s,t,v,w,x,y \mid yv = wy,\ tyt^{-1}x = wtyt^{-1},\ x^2ty^{-1}t^{-1}ys^{-1}w^{-1}xv^{-1} = 1,$$

$$sv = ts,\ vs = xv,\ wt = tw,\ xs = tx\rangle.$$

Setting $s = x\alpha$, $t = x\beta$, $v = x\gamma$ and $w = x\delta$, we obtain the presentation

$$\langle \alpha,\beta,\gamma,\delta,x,y \mid [x,y] = xy\gamma(xy)^{-1}.x\delta x^{-1},\ \beta.y\beta y^{-1} = \delta.x\beta x^{-1}.xy\beta^{-1}(xy)^{-1}.[x,y],$$

$$x^2\beta x^{-2}.x^2y^{-1}\beta^{-1}yx^{-2} = \gamma\delta.x\alpha x^{-1}.xy^{-1}[x,y]^{-1}yx^{-1}$$

$$\delta x\beta = \beta x\delta,\ \alpha x\gamma = \beta x\alpha,\ \gamma x\alpha = x\gamma,\ x\alpha = \beta x\rangle$$

in which α, β, γ and δ represent elements of π', which is the normal closure of the images of these generators. The subquotient π'/π'' is generated as a module over $\mathbb{Z}[\pi/\pi'] \cong \Lambda_2 = \mathbb{Z}[x^\pm, y^\pm]$ by the images of γ and δ, with the relations

$$(x+1)(y-1)(x-1)[\gamma] = xy[\gamma] - x[\delta],$$

$$(x-1)^2[\gamma] = (x-1)[\delta],$$

and

$$(x^2 - x + 1)[\gamma] = 0,$$

since $[\alpha] = x^{-1}(x-1)[\gamma]$ and $[\beta] = (x-1)[\gamma]$. Adding the first two relations and rearranging gives

$$[\delta] = -((x^2 - x + 1)y + 2 - 2x)[\gamma] = 2(x-1)[\gamma].$$

Hence $\pi'/\pi'' \cong \Lambda_2/(x^2 - x + 1, 3(x-1)^2) = \Lambda_2/(x^2 - x + 1, 3)$. As a module over the subring $\mathbb{Z}[x, x^{-1}]$, this is infinitely generated, but as a module over $\mathbb{Z}[y, y^{-1}]$ it has two generators. Therefore, there is no automorphism of π which induces an isomorphism $\mathrm{Ker}(j_{X*}) = \pi' \rtimes \langle x \rangle \cong \mathrm{Ker}(j_{Y*}) = \pi' \rtimes \langle y \rangle$. Hence (X, M) and (Y, M) are not homotopy equivalent as pairs, although $X \simeq Y$.

Does M have any other abelian embeddings with neither complementary component homeomorphic to X, perhaps corresponding to distinct link presentations? Is this 3-manifold homeomorphic to a 3-manifold $M(\tilde{L})$ via a homeomorphism which does not preserve the meridians?

There is just one 3-manifold with π elementary amenable and $\beta = 2$ which embeds in S^4. This is the $\mathbb{N}il^3$-manifold $M = M(Wh)$ and $\pi \cong F(2)/\gamma_3 F(2)$. The embedding j_{Wh} is abelian, since the components of Wh are unknotted, but the complementary regions are not aspherical and so Lemma 7.9 does not apply. All epimorphisms from π to $\mathbb{Z}$ are equivalent under composition with automorphisms, and each automorphism of π is induced by a self-diffeomorphism of M, since M is Haken. Thus, if $j : M \to S^4$ is another abelian embedding and we fix a homotopy equivalence $h : X \to X(j_{Wh})$ then we may assume that $\pi_1(h \circ j_X) = j_{X*}$. Can we choose h to be a homotopy equivalence of pairs, *rel* M?

We have a more general result, albeit rather weak.

THEOREM 8.16. *Let M be a 3-manifold with fundamental group π such that $\pi/\pi' \cong \mathbb{Z}^2$. Then there are at most four abelian embeddings of M with given homotopy types for the pairs (X, M) and (Y, M).*

PROOF. Let W be a 4-manifold with connected boundary M and such that $W \simeq S^1 \vee S^2$. Then $\mathbb{Z}[\pi_1 W] \cong \Lambda$, and $L_5(\Lambda)$ acts trivially on $\mathcal{S}_{TOP}(W, M)$, by [**FMGK**, Theorem 6.7]. Since $H_2(W; \mathbb{F}_2) = \mathbb{Z}/2\mathbb{Z}$, there are at most two possibilities for each complementary region realizing the given homotopy types. □

8.5. The Higher Rank Cases

Theorems 8.8 and 8.13 have analogues when $\beta = 3$, 4 or 6.

THEOREM 8.17. *Let M be a 3-manifold with fundamental group π such that $\pi/\pi' \cong \mathbb{Z}^3$. If $j : M \to S^4$ is an abelian embedding then $X \simeq T$ and $Y \simeq S^1 \vee 2S^2$, while $H_1(M;\mathbb{Z}[\pi_X]) \cong \mathbb{Z}$ and $H_1(M;\mathbb{Z}[\pi_Y])$ is a torsion $\mathbb{Z}[\pi_Y]$-module of projective dimension 1 and which can be generated by two elements. The component X is determined up to homeomorphism by M and j_{X*}.*

PROOF. The classifying map $c_X : X \to K(\pi_X, 1) \simeq T$ is a homotopy equivalence, by Theorem 7.7, since $c.d.X = c.d.T = 2$ and $\chi(X) = \chi(T) = 0$. The equivalence $Y \simeq S^1 \vee 2S^2$ follows from Corollary 7.7.1, since $\pi_Y \cong \mathbb{Z}$ and $\chi(Y) = 2$.

Since $H_2(X;\mathbb{Z}[\pi_X]) = 0$, the exact sequence of homology for the pair (X, M) with coefficients $\mathbb{Z}[\pi_X]$ reduces to an isomorphism $H_2(X, M;\mathbb{Z}[\pi_X]) \cong H_1(M;\mathbb{Z}[\pi_X])$, and so $H_1(M;\mathbb{Z}[\pi_X]) \cong \overline{H^2(X;\mathbb{Z}[\pi_X])} \cong \mathbb{Z}$, by Poincaré duality.

Similarly, there is an exact sequence

$$0 \to \mathbb{Z} \to H_2(M;\mathbb{Z}[\pi_Y]) \to \mathbb{Z}[\pi_Y]^2 \to \mathbb{Z}[\pi_Y]^2 \to H_1(M;\mathbb{Z}[\pi_Y]) \to 0,$$

since $H_2(Y;\mathbb{Z}[\pi_Y]) \cong \mathbb{Z}[\pi_Y]^2$ and $H_2(Y, M;\mathbb{Z}[\pi_Y]) \cong \overline{H^2(Y;\mathbb{Z}[\pi_Y])}$. Let $A = \pi'/\pi''$, considered as a $\mathbb{Z}[\pi/\pi']$-module. Then A is finitely generated as a module, since $\mathbb{Z}[\pi/\pi']$ is a noetherian ring. Let $\{x, y, z\}$ be a basis for π/π' such that $j_{X*}(y) = 0$ and $j_{Y*}(x) = j_{Y*}(z) = 0$. Then $H_1(M;\mathbb{Z}[\pi_X]) \cong \mathbb{Z}$ is an extension of $\mathbb{Z}$ by $A/(y-1)A$, and so $A = (y-1)A$. Similarly, $H_1(M;\mathbb{Z}[\pi_Y])$ is an extension of $\mathbb{Z}^2$ by $A/(x-1, z-1)A$. Together these observations imply that $H_1(M;\mathbb{Z}[\pi_Y])$ is a torsion $\mathbb{Z}[\pi_Y]$-module, and so the fourth homomorphism in the above sequence is a monomorphism. Thus, $H_1(M;\mathbb{Z}[\pi_Y])$ is a torsion $\mathbb{Z}[\pi_Y]$-module with projective dimension $\leqslant 1$, and is clearly generated by two elements. (Note also that a torsion $\mathbb{Z}[\pi_Y]$-module of projective dimension 0 is 0.)

Since X is aspherical, Lemma 7.9 applies, and so the homotopy type of the pair (X, M) is determined by M and the homomorphism j_{X*}. The final assertion follows (as in Lemma 6.10), since $L_5(\mathbb{Z}[\pi_X])$ acts trivially on $\mathcal{S}_{TOP}(X, M)$, by [**FMGK**, Theorem 6.7], and the integral Novikov Conjecture holds for $\pi_X \cong \mathbb{Z}^2$. □

Since $L_5(\mathbb{Z}[\pi_Y])$ also acts trivially on $\mathcal{S}_{TOP}(Y, M)$, by [**FMGK**, Theorem 6.7], and $H_2(Y;\mathbb{F}_2) \cong (\mathbb{Z}/2\mathbb{Z})^2$, there are at most four possibilities for Y, given the homotopy type of (Y, M).

The link $L = 9^3_{21}$ has a unique partition as a bipartedly slice link, and the fundamental groups of the corresponding complementary regions are $\pi_X \cong F(2)$ and $\pi_Y \cong \mathbb{Z}$. Then $M = M(9^3_{21}) \cong (S^2 \times S^1)\#M(5^2_1)$, so $\pi \cong \mathbb{Z} * F(2)/\gamma_3 F(2)$, with presentation $\langle x, y, z \mid [x, y] \leftrightharpoons x, y\rangle$. It is not hard to show that the kernel of any epimorphism $\phi : \pi \to \langle t\rangle \cong \mathbb{Z}$ has rank $\geqslant 1$ as a $\mathbb{Z}[t, t^{-1}]$-module. Hence M has no abelian embedding, by Theorem 8.17.

The 3-torus $T^3 = \mathbb{R}^3/\mathbb{Z}^3$ has an abelian embedding, as the boundary of a regular neighbourhood of an unknotted embedding of T in S^4. This manifold may be obtained by 0-framed surgery on the Borromean rings Bo, and also on 9^3_{18}. The three bipartite partitions of Bo lead to equivalent embeddings. (However, these are clearly not isotopic!) The link 9^3_{18} has two bipartedly slice partitions (both bipartedly trivial). Any such embedding of T^3 has $X \cong T \times D^2$ and $Y \simeq S^1 \vee 2S^2$. Does T^3 have an essentially unique abelian embedding?

If M is Seifert fibred and $\pi/\pi' \cong \mathbb{Z}^3$ then it has generalized Euler invariant $\varepsilon = 0$, and so is a mapping torus $T_g \rtimes_\theta S^1$, with orientable base orbifold and monodromy θ of finite order. Are there any such manifolds other than the 3-torus which have abelian embeddings?

Suppose that $\beta = 3$ and M has an embedding j such that $H_1(Y;\mathbb{Z}) = 0$. If $f : \pi \to \mathbb{Z}^2$ is an epimorphism with kernel κ and $R = \mathbb{Z}[\pi/\kappa]_{f(S)}$ then $H_1(M;R)$ has rank 2, by Lemma 8.5, and so the condition of Theorem 8.17 does not hold. Therefore, no such 3-manifold can also have an abelian embedding.

THEOREM 8.18. *Let M be a 3-manifold with fundamental group π such that $\pi/\pi' \cong \mathbb{Z}^4$. If $j : M \to S^4$ is an abelian embedding then $X \simeq Y \simeq T \vee S^2$. Hence $H_1(M;\mathbb{Z}[\pi_X])$ is a quotient of $\mathbb{Z}[\pi_X] \oplus \mathbb{Z}$ by a cyclic submodule (and similarly for $H_1(M;\mathbb{Z}[\pi_Y])$).*

PROOF. As in Corollary 7.7.1, generators for $\pi_X \cong \mathbb{Z}^2$ and $\pi_2 X \cong \mathbb{Z}[\pi_X]$ determine a map from $T^{[1]} \vee S^2$ to X. This extends to a 2-connected map from $T \vee S^2$ to X, which is a homotopy equivalence by Theorem 7.7. Hence $X \simeq T \vee S^2$.

The second assertion follows from the exact sequence of homology for (X, M) with coefficients $\mathbb{Z}[\pi_X]$, since $H^2(X;\mathbb{Z}[\pi_X]) \cong \mathbb{Z}[\pi_X] \oplus \mathbb{Z}$. Parallel arguments apply for Y and $H_1(M;\mathbb{Z}[\pi_Y])$. □

An orientation for $V = \mathbb{Z}^4$ determines an isomorphism $\wedge^3 V \cong Hom(V,\mathbb{Z})$, and it follows easily that all alternating 3-forms $\mu : V \to \mathbb{Z}$ with the same set of values are equivalent under the action of $Aut(V) \cong GL(4;\mathbb{Z})$. If M has an abelian embedding and $\beta = 4$, we may refine Corollary 7.3.1(2) to show that μ_M is surjective. (This uses Poincaré duality and the fact that $X \simeq Y \simeq T \vee S^2$.)

The argument below for the final case ($\beta = 6$) is adapted from Wall's proof that the $(n-1)$-skeleton of a PD_n-complex is essentially unique [**Wa67**, Theorem 2.4].

THEOREM 8.19. *Let M be a 3-manifold with fundamental group π such that $\pi/\pi' \cong \mathbb{Z}^6$. If $j : M \to S^4$ is an abelian embedding then $X \simeq Y \simeq T^{3[2]}$, the 2-skeleton of the 3-torus T^3, while $H_1(M;\mathbb{Z}[\pi_X])$ and $H_1(M;\mathbb{Z}[\pi_Y])$ are cyclic $\mathbb{Z}[\pi_X]$- and $\mathbb{Z}[\pi_Y]$-modules (respectively) of projective dimension $\leqslant 1$.*

PROOF. Since $\beta = 6$ and j is abelian, we may identify π_X with $\mathbb{Z}^3$. The first part of the second paragraph of Theorem 8.8 applies to show that $\pi_2 X$ is isomorphic to $\Lambda_3 = \mathbb{Z}[\mathbb{Z}^3]$. Let C_* and D_* be the equivariant chain complexes of the universal covers of $T^{3[2]}$ and X, respectively. Since these are partial resolutions of $\mathbb{Z}$ there is a chain map $f_* : C_* \to D_*$ such that $H_0(f)$ is an isomorphism. Clearly $H_1(f)$ is also an isomorphism. We shall modify our choice of f_* so that it is a chain homotopy equivalence.

The Λ_3-modules $H_2(C_*) < C_2$ and $H_2(D_*) < D_2$ are free of rank 1. Let $t \in C_2$ and $x \in D_2$ represent generators for these submodules, and let t^* and x^* be the Kronecker dual generators of the cohomology modules $H^2(C^*) = Hom(H_2(C_*),\Lambda_3) \cong \Lambda_3$ and $H^2(D^*) = Hom(H_2(D_*),\Lambda_3) \cong \Lambda_3$, respectively. Let $f_i' = f_i$ for $i = 0, 1$, and let $f_2'(u) = f_2(u) - z^*(u)x$ for all $u \in C_2$, where $z^* = H^2(f^*)(x^*) - t^* \in$

$Hom(H_2(C_*), \Lambda_3)$. Then f'_* is again a chain homomorphism, and $H_2(f'_*)$ is an isomorphism. Hence f_* is a chain homotopy equivalence. This may be realized by a map from $T^{3[2]}$ to the 2-skeleton $X^{[2]}$, and the composite with the inclusion $X^{[2]} \subseteq X$ is then a homotopy equivalence.

A similar argument applies for Y. The second assertion follows as before. □

We may also refine Corollary 7.3.1(2) to show that if M has an abelian embedding and $\beta = 6$ then μ_M is surjective. In this case there may be a further constraint, since even after extending coefficients to $\mathbb{R}$ there are five orbits of non-zero 3-forms on $\mathbb{R}^6$ under the action of $GL(6, \mathbb{R})$ [**Bry06**].

Lemma 8.5 and Theorems 8.18 and 8.19 again imply that when $\beta = 4$ or 6 no 3-manifold which has an embedding j such that $H_1(Y) = 0$ can also have an abelian embedding. However, if L is the 4-component link obtained from Bo by adjoining a parallel copy of one component, then $M(L)$ has an abelian embedding with $X \cong Y$ and $\chi(X) = 1$. It also has an embedding with $\chi(X) = -1$. We shall not give more details, as no natural examples demand our attention in these cases.

8.6. 2-Component Links with $\ell \neq 0$

If M is a rational homology sphere with an abelian embedding then $\pi/\pi' \cong (\mathbb{Z}/\ell\mathbb{Z})^2$ and $\pi_X \cong \pi_Y \cong \mathbb{Z}/\ell\mathbb{Z}$, for some $\ell > 0$. In particular, if L is a 2-component link with linking number $\ell \neq 0$ then $M(L)$ is a rational homology sphere. If the components of L are unknotted then L determines an embedding j_L, which is clearly abelian. There is again a necessary condition for the existence of an abelian embedding.

LEMMA 8.20. *Let M be a 3-manifold with fundamental group π such that $\pi/\pi' \cong (\mathbb{Z}/\ell\mathbb{Z})^2$, for some $\ell > 0$. If $j : M \to S^4$ is an abelian embedding then $X \simeq Y \simeq P_\ell$, and $H_1(M; \mathbb{Z}[\pi_X])$ and $H_1(M; \mathbb{Z}[\pi_Y])$ are cyclic $\mathbb{Z}[\pi_X]$- and $\mathbb{Z}[\pi_Y]$-modules (respectively) and are quotients of $\mathbb{Z}^{\ell-1}$, as abelian groups.*

PROOF. The first assertion holds by Lemma 8.4. The second part then follows from the exact sequences of homology for the universal covering spaces of the pairs (X, M) and (Y, M), since $\widetilde{X} \simeq \widetilde{Y} \simeq \vee^{\ell-1} S^2$. (Note that $H_2(\widetilde{X})$ and $H^2(\widetilde{X})$ are each isomorphic to the augmentation ideal of $\mathbb{Z}[\pi_X]$, which is cyclic as a module and free of rank $\ell - 1$ as an abelian group.) □

To use Theorem 8.20 to show that some M has no abelian embedding we must consider all possible bases for $Hom(\pi, \mathbb{Z}/\ell\mathbb{Z})$, or, equivalently, for π/π'.

Six of the eight rational homology 3-spheres with elementary amenable groups and which embed in S^4 have such link presentations, with $\ell \leqslant 4$. The simplest such links are the $(2, 2\ell)$-torus links. If L is such a link then $M(L) \cong M(0; (\ell, 1), (\ell, 1), (\ell, -1))$, while $X_L \cong Y_L$, since L is interchangeable. (In particular, the (2,2)-torus link 2^2_1 gives S^3.)

If $\ell = 1$ then M is an integral homology 3-sphere, and so has an unique abelian embedding, by Theorem 8.6.

When $\ell = 2$ we have $X \simeq Y \simeq \mathbb{RP}^2$, and the composite $\partial\widetilde{X} \subset \widetilde{X} \simeq S^2$ induces an isomorphism on H_2.

The quaternion manifold $M = S^3/Q(8) \cong M(4^2_1)$ has an essentially unique abelian embedding. The complementary regions are homeomorphic to the total space N of the disc bundle over $\mathbb{RP}^2$ with Euler number 2 [**Law84**]. (Lawson constructed a self-homotopy equivalence of N which is the identity outside a regular neighbourhood of an essential S^1, and which has non-trivial normal invariant. Thus, every element of $\mathcal{S}_{TOP}(N, \partial N)$ is represented by a homeomorphism. His construction extends to all $X \simeq \mathbb{RP}^2$. Do all the resulting self-homotopy equivalences have non-trivial normal invariant?

The links 9^2_{38}, 9^2_{57} and 9^2_{58} each have unknotted components, asymmetric Alexander polynomial and linking number 2.

Let L be the link obtained by tying a slice knot with non-trivial Alexander polynomial (such as the stevedore's knot 6_1) in one component of 4^2_1. Then $M(L)$ embeds in S^4, but does not satisfy Lemma 8.20, since for two of the three 2-fold covers of $M(L)$ the first homology is not cyclic as an abelian group. Hence $M(L)$ has no abelian embedding.

Suppose next that $\ell = 3$. The manifold $M(6^2_1)$ is a $\mathbb{N}il^3$-manifold with Seifert base the flat orbifold $S(3,3,3)$, and $X_L \cong Y_L$.

The most interesting example with $\ell = 4$ is perhaps $M(8^2_2)$, which is the $\mathbb{N}il^3$-manifold $M(-1;(2,1),(2,3))$. The link 8^2_2 is interchangeable, and so $X_L \cong Y_L$. Each of the links 9^2_{53} and 9^2_{61} has unknotted components and $\ell = 4$, and gives a $\mathbb{S}ol^3$-manifold with an abelian embedding. Is either of these links interchangeable?

In the range $\ell \leqslant 4$ the groups $Wh(\mathbb{Z}/\ell\mathbb{Z})$ and $L_5(\mathbb{Z}[\mathbb{Z}/\ell\mathbb{Z}])$ are each 0 [**Coh, Wa76**], while $H_2(X;\mathbb{F}_2) = H_2(Y;\mathbb{F}_2) = 0$ if ℓ is odd. The main difficulty in identifying abelian embeddings of these manifolds lies in identifying the homotopy types of the pairs (X, M) and (Y, M).

There remains one more $\mathbb{S}ol^3$-manifold which embeds in S^4. This is $M_{2,4}$, which arises from surgery on the link $L = (8_{20}, U)$ of Figure 5.1. The knot 8_{20} bounds a slice disc $D \subset D^4$ obtained by desingularizing the obvious ribbon disc of this figure. Then $M(L)$ has an embedding with one complementary region X obtained from $D^4 \subset S^4$ by deleting a regular neighbourhood of D and adding a 2-handle along the unknotted component U, and the other being $Y_L = \overline{S^4 \setminus X_L}$. The "ribbon group" $\pi_1(D^4 \setminus D)$ has a presentation obtained by adjoining the relation $z = x$ to the Wirtinger presentation for $\pi 8_{20}$ [**AIL**, Theorem 1.15]. It is easily seen that this presentation reduces to $\langle y, z \mid yzy = zyz\rangle$. Adding a 2-handle along the unknotted component kills the a-longitude $\ell_a = zyzyzy^{-1} = zyyzyy^{-1} = zy^2z$, and so we obtain the presentation $\langle y, z \mid yzy = zyz,\ y^2z^2 = 1\rangle$. The corresponding group is the semi-direct product $\mathbb{Z}/3\mathbb{Z} \rtimes_{-1} \mathbb{Z}/4\mathbb{Z}$. On the other hand, $\pi_{Y_L} \cong \mathbb{Z}/4\mathbb{Z}$.

The group $\pi = \pi_1 M(L)$ has the presentation

$$\langle x, y, u \mid xyx^{-1} = y^{-1},\ ux^2y^{-4}u^{-1} = x^{-2}y^4,\ u^2 = x^4y^{-9}\rangle.$$

(See Section 5.2) The abelianization π^{ab} is generated by the images of x and u. Let $\lambda_{i,j} : \pi \to \mathbb{Z}/4\mathbb{Z}$ be the epimorphism sending x, u to $i, j \in \mathbb{Z}/4\mathbb{Z}$, respectively. It can be shown that the abelianization of $\mathrm{Ker}(\lambda_{i,j})$ is a quotient of the augmentation ideal in $\mathbb{Z}[\mathbb{Z}/4\mathbb{Z}]$, for $(i,j) = (1,0)$ or (2,1). Since these epimorphisms form a basis for $Hom(\pi, \mathbb{Z}/4\mathbb{Z})$, we cannot use Lemma 8.20 to rule out an abelian embedding for $M_{2,4}$. Is there a 2-component link with unknotted components which gives rise to this manifold?

8.7. Some Remarks on the Mixed Cases

If M has an abelian embedding such that $\pi_X \cong G_k = \mathbb{Z} \oplus (\mathbb{Z}/k\mathbb{Z})$, for some $k > 1$, then $\chi(X) = 1$, by Lemma 7.11. Hence $\chi(Y) = 1$ and so $\pi_Y \cong G_k$ also. Therefore $H_1(M) \cong \mathbb{Z}^2 \oplus (\mathbb{Z}/k\mathbb{Z})^2$, which requires four generators. The simplest examples may be constructed from 4-component links obtained by replacing one component of the Borromean rings Bo by its $(2k, 2)$ cable.

In this case even the determination of the homotopy types of the complements is not clear. The group G_k has minimal presentations

$$\mathcal{P}_{k,n} = \langle a, t \mid a^k,\ ta^n = a^n t\rangle,$$

where $0 < n < k$ and $(n, k) = 1$. The 2-complexes $S_{k,n} = S^1 \vee P_k \cup_{[t,a^n]} e^2$ associated to these presentations have Euler characteristic 1, and it is easy to see that there are maps between them which induce isomorphisms on fundamental groups. We may identify $S_{k,n}$ with $T \cup MC \cup P_k$, where MC is the mapping cylinder of the degree-n map $z \mapsto z^n$ from $\{1\} \times S^1 \subset T$ to the 1-skeleton $S^1 \subset P_k$. In particular, $S_k = S_{k,1} = T \cup_{a^k} e^2$ is the 2-skeleton of $S^1 \times P_k$. From these descriptions it is easy to see that (1) automorphisms of G_k which fix the torsion subgroup $A = \langle a \rangle$ may be realized by self-homeomorphisms of $S_{k,n}$ which act by reflections and Dehn twists on T, and fix the second 2-cell; and (2) the automorphism which fixes t and inverts a is induced by an involution of $S_{k,n}$.

Let $C(k, n)_*$ be the cellular chain complex of the universal cover of $S_{k,n}$. A choice of basepoint for $S_{n,k}$ determines lifts of the cells of $S_{k,n}$, and hence isomorphisms $C(k, n)_0 \cong \Gamma$, $C(k, n)_1 \cong \Gamma^2$ and $C(k, n)_2 \cong \Gamma^2$. The differentials are given by $\partial_1 = (a - 1, t - 1)$ and $\partial_2^n = \begin{pmatrix} (t-1)\nu_n & \rho \\ 1-a & 0 \end{pmatrix}$, where $\nu_n = \Sigma_{0 \leqslant i < n} a^i$ and $\rho = \nu_k$. Let $\{e_1, e_2\}$ be the standard basis for $C(k, n)_2$. Then $\Pi_{k,n} = \pi_2 S_{k,n} = \mathrm{Ker}(\partial_2^n)$ is generated by $g = \rho e_1 - n(t-1)e_2$ and $h = (a-1)e_2$, with relations $(a-1)g = n(t-1)h$ and $\rho h = 0$. It can be shown that $\Pi_{k,n} \cong \alpha^* \Pi_{k,m}$, where α is the automorphism of G_k such that $\alpha(t) = t$ and $\alpha(a) = a^r$, where $n \equiv rm \ mod\ k$. Is there a chain homotopy equivalence $C(k, n)_* \simeq \alpha^* C(k, m)_*$?

Is every finite 2-complex S with $\pi_1 S \cong G_k$ and $\chi(S) = 1$ homotopy equivalent to $S_{k,n}$, for some n? The key invariants are the Γ-module $\pi_2 S$ and the k-invariant in $H^3(G_k; \pi_2 S)$. Let $S_{\langle t \rangle}$ be the finite covering space with fundamental group $\langle t \rangle \cong \mathbb{Z}$. If M is a finitely generated submodule of a free Γ-module then $H^i(\langle t \rangle; M) = 0$ for $i \neq 1$, while $H^1(\langle t \rangle; M) = M_t = M/(t-1)M$. Hence the spectral sequence

$$H^p(A; (H^q(\langle t \rangle; M)) \Rightarrow H^{p+q}(G_k; M)$$

collapses, to give $H^{p+1}(G_k; M) \cong H^p(A; M_t)$. If $M = \pi_2 S$ then $M_t \cong H_2(S_{\langle t \rangle}; \mathbb{Z})$, as a $\mathbb{Z}[A]$-module. When $M = \Pi_{k,n}$ it is easy to see that $M_t \cong \mathbb{Z} \oplus I_A$, where I_A is the augmentation ideal of $\mathbb{Z}[A]$, and so $H^2(A; M_t) \cong H^2(A; \mathbb{Z}) \cong \mathbb{Z}/k\mathbb{Z}$.

Let V and W be finite 2-complexes with $\pi_1 V \cong \pi_1 W \cong G_k$, and let $\Gamma = \mathbb{Z}[G_k]$. Then $\chi(V) \geqslant 1$ and $\chi(W) \geqslant 1$, and an application of Schanuel's Lemma to the chain complexes of the universal covers gives

$$\pi_2 V \oplus \Gamma^{\chi(W)+n} \cong \pi_2 W \oplus \Gamma^{\chi(V)+n},$$

for $n >> 0$. Taking $W = S_{k,1}$, we see that $H^3(G;\pi_2 V) \cong \mathbb{Z}/k\mathbb{Z}$, for all such V.

Even if we can determine the homotopy types of the 2-complexes S with $\pi_1 S$ and $\chi(S) = 1$, and the homotopy types of the pairs (X, M) for a given M, the groups $L_5^s(\mathbb{Z}[G])$ are commensurable with $L_4(\mathbb{Z}[\mathbb{Z}/k\mathbb{Z}])$, which has rank $\lfloor \frac{k+1}{2} \rfloor$, and so characterizing such abelian embeddings up to isotopy may be difficult.

The S^1-bundle spaces $M(-2;(1,0))$ (the half-turn flat 3-manifold G_2), and $M(-2;(1,4))$ (a $\mathbb{N}il^3$-manifold) are the total spaces of S^1-bundles over Kb, and $\beta = 1$ for these manifolds. They do not have abelian embeddings, since π/π' has non-trivial torsion. In each case π requires three generators, and so they cannot be obtained by surgery on a 2-component link. However, they may be obtained by 0-framed surgery on the links 8^3_9 and 9^3_{19}, respectively. For the embeddings defined by these links, $X \simeq Kb$ and $\pi_Y = \mathbb{Z}/2\mathbb{Z}$. Since X is aspherical we may apply Lemma 7.9 to show that in each case the pair (X, M) is homotopy equivalent *rel* ∂ to the corresponding disc bundle space. The group $L_5(\mathbb{Z}[\mathbb{Z} \rtimes_{-1} \mathbb{Z}])$ acts trivially on the structure set $\mathcal{S}_{TOP}(X, \partial X)$, by [**FMGK**, Theorem 6.7]. Thus there are only finitely many possible homeomorphism types for X. As in Lemma 8.4, Y is homotopy equivalent to a finite 2-complex, and hence $Y \simeq \mathbb{RP}^2 \vee S^2$. However, we do not yet know how to identify the pair (Y, M).

It is easy to find 3-component bipartedly trivial links L such that X_L is aspherical and $\pi_{X_L} \cong BS(1,m)$, for $m \neq 0$. (If $m \neq 1$ then $H_2(X) = 0$, and if $m = 2$ then M is a homology handle.) The group $L_5(\mathbb{Z}[\pi_X])$ acts trivially on $\mathcal{S}_{TOP}(X, M)$ [**FMGK**, Lemma 6.9]. If m is even then $H_2(X;\mathbb{F}_2) = 0$, and so X is determined up to homeomorphism by M and the homomorphism j_{X*}. (See also [**DH25**].) In this case Y is homotopy equivalent to a finite 2-complex, by Lemma 8.4, since $\pi_Y \cong \mathbb{Z}/(m-1)\mathbb{Z}$, $\chi(Y) = 2$ and $c.d.Y \leqslant 2$. Hence $Y \simeq P_{m-1} \vee S^2$ [**DS73**]. However, we do not yet know how to identify the pair (Y, M).

CHAPTER 9

Nilpotent Embeddings

The broader class of nilpotent groups is of particular interest. If π_X is nilpotent then $j_{X*} = \pi_1 j_X$ is onto, since $H_1(j_X)$ is onto, and any subset of a nilpotent group G whose image generates the abelianization G/G' generates G. Since j_{X*} is onto, $c.d.X \leqslant 2$, by Theorem 7.10. (Similarly, if π_Y is nilpotent then j_{Y*} is onto and $c.d.Y \leqslant 2$.) If π_X and π_Y are each nilpotent then j is bi-epic, by Lemma 2.11.

There are also purely algebraic reasons why nilpotent groups should be of interest. Firstly, there is the well-known connection between homology, lower central series and (Massey) products [**Dw75, St65**]. Secondly, if a group G is finite or solvable and every homomorphism $f : H \to G$ which induces an epimorphism on abelianization is an epimorphism, then G must be nilpotent. (See pages 132 and 460 of [**Rob**].) However, even the class of 2-generator nilpotent groups with balanced presentations is not known. (We expect that nilpotent groups of large Hirsch length should have negative deficiency, and so should not arise in this context.)

We begin by showing that if G is a finitely generated nilpotent group other than $\mathbb{Z}$ or $\mathbb{Z}^2$ then $\beta_2(G;F) \geqslant \beta_1(G;F)$ for some prime field F, and we summarize what is presently known about (infinite) homologically balanced nilpotent groups. (See Appendix B for more details.) We then give strong constraints on nilpotent groups arising in our context, parallel to Theorem 8.1. Several examples of abelian and nilpotent embeddings are given in Section 9.3, and in the final section we show that if M has a restrained embedding determined by a bipartedly trivial link then $\beta \leqslant 8$.

9.1. Wang Sequence Estimates

An automorphism α of an abelian group A is *unipotent* if $\alpha - id_A$ is nilpotent. The following lemma is a particular case of a result of P. Hall [**Rob**, 5.2.1].

LEMMA (Hall). *Let ψ be an automorphism of a finitely generated nilpotent group N. Then $G = N \rtimes_\psi \mathbb{Z}$ is nilpotent if and only if ψ^{ab} is unipotent.* □

We shall extend the term "unipotent", to say that an automorphism ψ of a finitely generated nilpotent group is unipotent if ψ^{ab} is unipotent. Our next lemma is probably known, but we have not found a published proof.

LEMMA 9.1. *Let ψ be a unipotent automorphism of a finitely generated nilpotent group N. Then $H_i(\psi;R)$ and $H^i(\psi;R)$ are unipotent, for all simple coefficients R and $i \geqslant 0$.*

PROOF. If N is cyclic then the result is clear. In general, N has a composition series with cyclic subquotients $\mathbb{Z}/p\mathbb{Z}$, where $p = 0$ or is prime. We shall induct on the number of terms in such a composition series. If N is infinite then ψ acts unipotently on $Hom(N, \mathbb{Z})$ and so fixes an epimorphism to $\mathbb{Z}$; if N is finite then ψ fixes an epimorphism to $\mathbb{Z}/p\mathbb{Z}$, for any p dividing the order of N.

Let K be the kernel of such an epimorphism. Then $\psi(K) = K$, by the choice of ψ; let $\psi_K = \psi|_K$. This is a unipotent automorphism of K [**Rob**, 5.2.10]. Hence the induced action of ψ on $H_i(K; R)$ is unipotent, for all i, by the inductive hypothesis. Let $\Lambda = \mathbb{Z}[N/K]$ and let B be a Λ-module. Then $H_i(N/K; B) = Tor_i^\Lambda(\mathbb{Z}, B)$ may be computed from the tensor product $C_* \otimes_\mathbb{Z} B$, where C_* is a resolution of the augmentation Λ-module $\mathbb{Z}$. If $B = H_i(K; R)$ then the diagonal action of ψ on each term of $C_* \otimes_\mathbb{Z} B$ is unipotent. The result is now a straightforward consequence of the Lyndon-Hochschild-Serre spectral sequence for N as an extension of N/K by K.

The argument for cohomology is similar. □

In fact we only need this lemma in degrees $\leqslant 2$. We shall usually assume that the coefficient ring is a field, and then homology and cohomology are linear duals of each other. Homology has an advantage deriving from the isomorphism $G^{ab} \cong H_1(G)$, but it is often more convenient to use cohomology instead.

If G is a finitely generated infinite nilpotent group then there is an epimorphism $f : G \to \mathbb{Z}$, and so $G \cong K \rtimes_\psi \mathbb{Z}$, where ψ is an automorphism of $K = \mathrm{Ker}(f)$ determined by conjugation in G. The homology groups $H_i(K; R) = H_i(G; R[G/K])$ are $R[G/K]$-modules, with a generator t of $G/K \cong \mathbb{Z}$ acting via $H_i(\psi; R)$. The (homology) Wang sequence for G (as an extension of $\mathbb{Z}$ by K) has the form

$$H_2(K; R) \xrightarrow{H_2(\psi;R)-I} H_2(K; R) \to H_2(G; R) \to$$

$$\to H_1(K; R) \xrightarrow{H_1(\psi;R)-I} H_1(K; R) \to H_1(G; R) \to R \to 0.$$

There is a similar Wang sequence for cohomology. (These are special cases of the Lyndon-Hochschild-Serre spectral sequences for the homology and cohomology with coefficients R of G as an extension of $\mathbb{Z}$ by K.)

LEMMA 9.2. *Let $G \cong K \rtimes_\psi \mathbb{Z}$ be a finitely generated nilpotent group, and let F be a field. Then*

(1) $\dim_F \mathrm{Cok}(H_2(\psi; F) - I) = \dim_F \mathrm{Ker}(H^2(\psi; F) - I) = \beta_2(G) - \beta_1(G) + 1$, *and so $\beta_2(G; F) \geqslant \beta_1(G; F) - 1$, with equality if and only if $\beta_2(K; F) = 0$;*
(2) *if $\beta_2(G; F) = \beta_1(G; F)$ then $H_2(K; F)$ is cyclic as a $F[G/K]$-module;*
(3) *$\beta_1(G; F) = 1 \Leftrightarrow \beta_2(G; F) = 0$, and then K is finite, $\beta_1(K; F) = 0$, and $h(G) = 1$;*
(4) *if $H_2(G; \mathbb{Z}) = 0$ then $G \cong \mathbb{Z}$.*

PROOF. Part (1) follows from the Wang sequences for the homology and cohomology of G as an extension of $\mathbb{Z}$ by K. The endomorphisms $H_i(\psi; F) - I$ have non-trivial kernel and cokernel if $H_i(K; F) \neq 0$, since they are nilpotent.

The $F[G/K]$-module $H = H_2(K;F)$ is finitely generated and is annihilated by a power of $t-1$, since $H_2(\psi;F)$ is unipotent. If $\beta_2(G;F) = \beta_1(G;F)$ then $\dim_F H/(t-1)H = 1$, by the exactness of the Wang sequence. Since $F[G/K] \cong F[t,t^{-1}]$ is a PID, it follows that H is cyclic as an $F[G/K]$-module.

Let $t \in G$ represent a generator of G/K. Then $F[G/K] \cong F[t,t^{-1}]$ is a PID and $H = H_2(K;F) = H_2(G;F[G/K])$ is a finitely generated $F[t,t^{-1}]$-module, with t acting via $H_2(\psi;F)$. This module is annihilated by a power of $t-1$, since $H_2(\psi;F)$ is unipotent. If $\beta_2(G;F) = \beta_1(G;F)$ then $\dim_F H/(t-1)H = 1$, by exactness of the Wang sequence. Since $F[G/K] \cong F[t,t^{-1}]$ is a PID, it follows that H is cyclic as an $F[G/K]$-module.

If $\beta_1(G;F) = 1$ then $H_1(K;F) = 0$, and so K is finite and $h(G) = 1$. Since K is finite it is the direct product of its Sylow subgroups, and the Sylow p-subgroup carries the p-primary homology of K. Hence if F has characteristic $p > 1$ and $H_1(K;F) = 0$ then the Sylow p-subgroup is trivial and $H_i(K;F) = 0$, for all $i \geqslant 1$. If F has characteristic 0 then $H_i(K;F) = 0$ for all $i \geqslant 1$ also. In each case, $H_i(G;F) = 0$, for all $i > 1$, and so $\beta_2(G;F) = 0$. Conversely, if $\beta_2(G;F) = 0$ then $H_1(\psi;F) - I$ is a monomorphism. Since $H_1(\psi;F) - I$ is nilpotent, $H_1(K;F) = 0$. Hence K is finite, so $h(G) = 1$, and $\beta_1(G;F) = 1$.

Part (4) is similar. If $H_2(G) = 0$ then $\psi^{ab} - I$ is a monomorphism, and so $K^{ab} = 0$. Hence $K = 1$ and $G \cong \mathbb{Z}$. □

In particular, if $h(G) = 1$ and T is the torsion subgroup of G then $\beta_1(T;\mathbb{F}_p) > 0$ if and only if $\beta_1(G;\mathbb{F}_p) > 1$. The fact that the torsion subgroup has non-trivial image in the abelianization does not extend to nilpotent groups G with $h(G) > 1$, as may be seen from the groups with presentation

$$\langle x, y \mid [x,[x,y]] = [y,[x,y]] = [x,y]^p = 1\rangle.$$

COROLLARY 9.2.1. *Let G be a finitely generated nilpotent group. Then*

(1) $\beta_2(G;\mathbb{Q}) < \beta_1(G;\mathbb{Q})$ *if and only if* $h(G) = 1$ *or* 2*;*
(2) *if* $\beta_2(G;\mathbb{F}_p) < \beta_1(G;\mathbb{F}_p)$ *for some prime p then G is infinite, G has no p-torsion and* $h(G) = \beta_1(G;\mathbb{F}_p) = 1$ *or* 2.

PROOF. If G is finite then $\beta_i(G;\mathbb{Q}) = 0$ for all $i > 0$, and if p divides the order of G then it follows from the Universal Coefficient Theorem that $\beta_2(G;\mathbb{F}_p) \geqslant \beta_1(G;\mathbb{F}_p)$, since ${}_pG^{ab}$ and G^{ab}/pG^{ab} have the same dimension.

Hence we may assume that G is infinite, and so $G \cong K \rtimes_\psi \mathbb{Z}$, where K is a finitely generated nilpotent group and ψ is a unipotent automorphism. Let F be a field. We may use Lemma 9.2 to show first that $\beta_2(K;F) = 0$ and then that $\beta_1(K;F) \leqslant 1$. Hence $\beta_1(G;F) \leqslant 2$.

If $F = \mathbb{Q}$ then either K is finite and $h(G) = 1$, or $h(K) = 1$ and $h(G) = 2$. The converse is clear, since G is then a finite extension of $\mathbb{Z}^{h(G)}$.

Suppose that $F = \mathbb{F}_p$ for some prime p. If $\beta_1(G;\mathbb{F}_p) = 1$ then K is finite, so $h(G) = 1$, and $\beta_2(K;\mathbb{F}_p) = 0$, so $\beta_1(K;\mathbb{F}_p) = 0$ and K has no p-torsion. If $\beta_1(G;\mathbb{F}_p) = 2$ then $\beta_1(K;\mathbb{F}_p) = 1$ and $\beta_2(K;\mathbb{F}_p) = 0$, so $h(K) = 1$ and K has no p-torsion. Hence $h(G) = 2$ and G has no p-torsion. □

It follows immediately that if $\beta_2(G;F) < \beta_1(G;F)$ for all prime fields F then $G \cong \mathbb{Z}$ or $\mathbb{Z}^2$.

The next result is a corrected version of Lemma 1 of **[Hi22]** (in which it was inadvertently assumed that $\beta_2(G) = \beta_1(G)$ if G is homologically balanced).

LEMMA 9.3. *Let G be a finitely generated nilpotent group and let $\beta = \beta_1(G;\mathbb{Q})$. Then G is homologically balanced if and only if $H_2(G)$ is a quotient of $\mathbb{Z}^\beta$; if $h(G) > 2$ then G is homologically balanced if and only if $H_2(G) \cong \mathbb{Z}^\beta$.*

PROOF. The first assertion follows from the Universal Coefficient exact sequences, since $G^{ab} \cong \mathbb{Z}^\beta \oplus B$, where B is finite. If $h(G) > 2$ then $\beta_2(G;\mathbb{Q}) \geqslant \beta$, by Corollary 9.2.1, and so $H_2(G)$ is a quotient of $\mathbb{Z}^\beta$ if and only if $H_2(G) \cong \mathbb{Z}^\beta$. □

In Chapter 1 we observed that a finite group T is homologically balanced if and only if $H_2(T) = 0$. It is not known whether every homologically balanced finite nilpotent group has a balanced presentation. The finite nilpotent 3-manifold groups $Q(8k) \times \mathbb{Z}/a\mathbb{Z}$ (with $(a, 2k) = 1$) have the balanced presentations

$$\langle x, y \mid x^{2ka} = y^2,\ yxy^{-1} = x^s\rangle,$$

where $s \equiv 1\ mod\ (a)$ and $s \equiv -1\ mod\ (2k)$. The other finite nilpotent groups F with 4-periodic cohomology (the generalized quaternionic groups $Q(2^n a, b, c) \times \mathbb{Z}/d\mathbb{Z}$, with a, b, c, d odd and pairwise relatively prime) have $H_2(F;\mathbb{Z}) = 0$, but we do not know whether they all have balanced presentations.

We shall summarize here some of the results of Appendix B on homologically balanced infinite nilpotent groups.

THEOREM. **(B.6)** *Let $G \cong T \rtimes_\psi \mathbb{Z}$, where T is a finite nilpotent group and ψ is a unipotent automorphism of T. If G is homologically balanced then*

1. *G can be generated by two elements;*
2. *if the Sylow p-subgroup of T is abelian then it is cyclic;*
3. *if T is abelian then $G \cong \mathbb{Z}/m\mathbb{Z} \rtimes_n \mathbb{Z}$, for some $m, n \neq 0$ such that m divides a power of $n - 1$.* □

Every semi-direct product $\mathbb{Z}/m\mathbb{Z} \rtimes_n \mathbb{Z}$ has a balanced presentation

$$\langle a, t \mid a^m = 1,\ tat^{-1} = a^n\rangle.$$

The simplest examples with T non-abelian are the groups $Q(8k) \rtimes \mathbb{Z}$, with the balanced presentations $\langle t, x, y \mid x^{2k} = y^2,\ tx = xt,\ tyt^{-1} = xy\rangle$, which simplify to

$$\langle t, y \mid [t, y]^{2k} = y^2,\ [t, [t, y]] = 1\rangle.$$

Let $m = p^s$, where p is a prime and $s \geqslant 1$, and let G be the group with presentation

$$\langle t, x, y \mid txt^{-1} = y,\ tyt^{-1} = x^{-1}y^2,\ yxy^{-1} = x^{m+1}\rangle.$$

If we conjugate the final relation with t to get the relation $x^{-1}yx = y^{m+1}$ then we see that the torsion subgroup T has presentation $\langle x, y \mid x^m = y^m,\ yxy^{-1} = x^{m+1}\rangle$. Moreover, G is nilpotent, $\zeta G = \langle x^m\rangle$ and $G' = \langle x^m, x^{-1}y\rangle$ is abelian. Hence G is metabelian. Each of the groups that we have described here is 2-generated and its torsion subgroup is homologically balanced.

It is not known whether there are infinitely many finitely generated, homologically balanced nilpotent groups G with $\beta_1(G) = 2$ and Hirsch length > 3. All known examples of homologically balanced nilpotent groups G with $h(G) > 1$ are torsion-free. If G is such a group then either $G \cong \mathbb{Z}^3$ or $\beta_1(G;\mathbb{Q}) \leqslant 2$ [**Hi20a**]. If, moreover, $h(G) \leqslant 5$ then G is metabelian, and is either free abelian of rank $\leqslant 3$ or a $\mathbb{N}il^3$-group Γ_q with presentation

$$\langle x, y, z \mid [x,y] = z^q,\ xz = zx,\ yz = zy\rangle,$$

for some $q \geqslant 1$, or is the $\mathbb{N}il^4$-group Ω with presentation

$$\langle t, u \mid [t,[t,[t,u]]] = [u,[t,u]] = 1\rangle.$$

9.2. Constraints on the Invariants

In this section we shall invoke Lubotzky's work on groups whose pro-p completions are p-adic analytic groups [**Lub83**]. This uses the notions of pro-p completion and of a p-good discrete group. A discrete group G is *p-good* if the canonical homomorphism from G to its pro-p completion $\widehat{G}_p$ induces isomorphisms on cohomology with coefficients $\mathbb{F}_p$. Virtually polycyclic groups are p-good for every prime p [**Ser**, Exercise 1.2.2], and their pro-p completions are p-adic analytic groups [**Lub83**, Proposition 2.6]. This class includes all finitely generated nilpotent groups.

THEOREM 9.4. *Let $j : M \to S^4$ be a bi-epic embedding such that π_X and π_Y are virtually polycyclic. Then*

(1) *if $\beta = \beta_1(M)$ is odd then X is aspherical, $\chi(X) = 0$ and $\pi_X \cong \mathbb{Z}$, $\mathbb{Z}^2$ or $\pi_1 Kb = BS(1,-1)$;*
(2) *if β is even then $\chi(X) = \chi(Y) = 1$ and $\beta = 0, 2, 4$ or 6.*

PROOF. If β is odd then $\chi(X) \leqslant 0$, by Lemma 2.2. On the other hand, $\beta_1^{(2)}(\pi_X) = 0$, since π_X is virtually polycyclic. Hence $\chi(X) = 0$ and X is aspherical, by Lemma 7.11. Since π_X is torsion-free and $H_i(\pi_X) = H_i(X) = 0$ for $i > 2$ the only possibilities are $\pi_X \cong \mathbb{Z}$, $\mathbb{Z}^2$ or $\pi_1 Kb = BS(1,-1)$, in which case $\beta_2(X) = 0, 1$ or 0, respectively. In particular, $\beta = 1$ or 3.

Suppose now that β is even. Then $0 < \chi(X) \leqslant \chi(Y)$ and $\chi(X) + \chi(Y) = 2$, so $\chi(X) = \chi(Y) = 1$. Hence π_X and π_Y are homologically balanced, by Lemma 2.2. Since $H_i(X;R) = 0$ for $i > 2$, we have $\beta_2(X;R) = \beta_1(X;R)$, and so $\beta_2(\pi_X;R) \leqslant \beta_1(\pi_X;R)$, for any coefficient ring R. Since π_X is virtually poly-$\mathbb{Z}$ it is p-good and is a pro-p analytic group. Hence $\beta_i(\widehat{\pi_X}_p;\mathbb{F}_p) = \beta_i(\pi_X;\mathbb{F}_p)$, for all i. Let $d = \beta_1(X;\mathbb{F}_p) = \beta_2(X;\mathbb{F}_p)$. Then $\widehat{\pi_X}_p$ has a balanced presentation (as a pro-p group) with d generators and d relators [**Ser**, Proposition I.27]. If $\beta > 2$ then $\widehat{\pi_X}_p \not\cong \widehat{\mathbb{Z}}_p$, and so $\beta_2(\pi_X;\mathbb{F}_p) > \frac{d^2}{4}$, by [**Lub83**, Theorem 2.7]. (A similar argument applies for π_Y.) Therefore $d \leqslant 3$ and $\beta \leqslant 2d \leqslant 6$. □

Nilpotent embeddings are always bi-epic, since homomorphisms to a nilpotent group which induce epimorphisms on abelianization are epimorphisms.

COROLLARY 9.4.1. *If j is nilpotent then*

(1) *if* β *is odd then either* $X \simeq S^1$ *and* $Y \simeq S^2$ *or* $X \simeq T$ *and* $Y \simeq S^1 \vee S^2 \vee S^2$.
(2) *if* β *is even then* $\chi(X) = \chi(Y) = 1$ *and* π_X *and* π_Y *can each be generated by three elements and are homologically balanced;*
(3) *if* $\beta = 6$ *then* $\pi_X \cong \pi_Y \cong \mathbb{Z}^3$.

PROOF. If β is odd then X is aspherical, and the embedding is abelian, by Theorem 9.4. The other details are given in Theorems 8.8 and 8.17.

If β is even then $\chi(X) = \chi(Y) = 1$, and so $\beta_2(\pi_X; F) \leqslant \beta_1(\pi_X; F)$ for all field coefficients F. Since π_X is finitely generated and nilpotent, there is a prime p such that π_X can be generated by $\beta_1(\pi_X; \mathbb{F}_p)$ elements, and $\beta_1(\pi_X; \mathbb{F}_p) \leqslant 3$, by Theorem 9.4. (Similarly for π_Y.)

If $\beta = 6$ then $\pi_X \cong \pi_Y \cong \mathbb{Z}^3$, by the main result of [**Hi22**]. □

If π_X is nilpotent and $\beta_1(X) = 0$ then π_X is finite, while if $\beta_1(X) = 1$ then $\pi_X \cong F \rtimes \mathbb{Z}$, where F is finite. Thus, if π_X and π_Y are torsion-free nilpotent and $\beta \leqslant 3$ then π_X and π_Y are abelian.

THEOREM 9.5. *If* M *has a nilpotent embedding and* $\beta \geqslant 3$ *then there is a non-zero Massey product of length at most* 4 *in* $H^2(M; \mathbb{Q})$.

PROOF. We may assume that π_X is non-abelian, and so $\beta = 4$, for if $\beta \geqslant 3$ and π_X is abelian there are non-zero cup products in $H^2(X; \mathbb{Q})$ and hence in $H^2(M; \mathbb{Q})$.

Hence $\beta_1(\pi_X) = 2$. Since π_X is nilpotent and homologically balanced $\pi_X/\gamma_5^{\mathbb{Q}}\pi_X$ is a proper quotient of $F(2)/\gamma_5^{\mathbb{Q}}F(2)$ [**FHT97**, Theorem 2], and so there is a non-zero Massey product of length at most 4 in $H^2(\pi_X; \mathbb{Q})$ [**Dw75**, Proposition 4.3]. Since $H^2(\pi_X; \mathbb{Q})$ maps injectively to $H^2(X; \mathbb{Q})$ and $H^2(M; \mathbb{Q})$, the theorem follows. □

In particular, $\#^\beta(S^2 \times S^1)$ has a nilpotent embedding if and only if $\beta \leqslant 2$.

THEOREM 9.6. *If* π_X *is nilpotent and* $H_1(Y)$ *is a non-trivial finite group then* π_X *is finite and* $\chi(X) = \chi(Y) = 1$. *If* $H_1(Y) = 0$ *then we may also have* $\pi_X \cong \mathbb{Z}$ *and* $\chi(X) = 0$.

PROOF. Since π_X is nilpotent, j_{X*} is an epimorphism, and so $c.d.X \leqslant 2$, by Theorem 7.10. Moreover, $\beta_1^{(2)}(\pi_X) = 0$ and so either $\chi(X) = 0$ and X is aspherical, or $\chi(X) = 1$, by Lemma 7.11. If $\chi(X) = 0$ then $\pi_X = 1$, $\mathbb{Z}$ or $\mathbb{Z}^2$, and $H_1(X)$ is torsion-free. Therefore, $H_1(Y) = 0$ (since $H_1(Y) = \tau_Y \cong \tau_X$), and so $H_2(X) = 0$.

If $\chi(X) = 1$ then $\chi(Y) = 1$, and so $H^1(X) \cong H_2(Y) = 0$ and $H_2(X) \cong H^1(Y) = 0$. Therefore, $H_1(X)$ is finite. Since π_X is nilpotent and has finite abelianization, it is finite. Moreover, $H_2(\pi_X) = H_2(\pi_Y) = 0$, since these groups are quotients of $H_2(X)$ and $H_2(Y)$, respectively. □

The complementary regions of the standard embedding of $S^2 \times S^1$ in S^4 are $X = D^3 \times S^1$ and $Y = S^2 \times D^2$, with fundamental groups $\mathbb{Z}$ and 1, respectively.

It is reasonable to restrict consideration further to torsion-free nilpotent groups, as such groups satisfy the Novikov Conjecture, and the surgery obstructions are maniable.

If G is torsion-free nilpotent of Hirsch length h then $c.d.G = h$. The first non-abelian examples are the $\mathbb{N}il^3$-groups Γ_q. Some of the argument of Theorem 8.19 for the group $\mathbb{Z}^3$ extends to the groups Γ_q. The homology of the pair (X, M) with coefficients $\mathbb{Z}[\pi_X]$ gives an exact sequence

$$H_2(X;\mathbb{Z}[\pi_X]) \to H^2(X;\mathbb{Z}[\pi_X]) \to H_1(M;\mathbb{Z}[\pi_X]) \to 0.$$

Let $K_X = \mathrm{Ker}(j_{X*}) = H_1(M;\mathbb{Z}[\pi_X])$ and $P = H_2(X;\mathbb{Z}[\pi_X])$. Since $c.d.X \leqslant 2$ and $c.d.\Gamma_q = 3$, an application of Schanuel's Lemma shows that P is a projective $\mathbb{Z}[\Gamma_q]$-module of rank 1. It is stably free since $\widetilde{K}_0(\mathbb{Z}[G]) = 0$ for torsion-free poly-$\mathbb{Z}$ groups G, and P has rank 1 since $\chi(X) = 1$. Since $Ext^i_{\mathbb{Z}[\pi_X])}(\mathbb{Z}, \mathbb{Z}[\pi_X]) = 0$ for $i \leqslant 2$ we then see that $H^2(X;\mathbb{Z}[\pi_X]) \cong P^\dagger = \overline{Hom_{\mathbb{Z}[\pi_X])}(P, \mathbb{Z}[\pi_X])}$, and so is also stably free of rank 1. We thus have an exact sequence

$$P \to P^\dagger \to K_X \to 0.$$

However, it is not clear that this is as potentially useful as the analogous conditions on abelian embeddings given in Chapter 8. Moreover, if G is a non-abelian poly-$\mathbb{Z}$ group then there are infinitely many isomorphism classes of stably free $\mathbb{Z}[G]$-modules P such that $P \oplus \mathbb{Z}[G] \cong \mathbb{Z}[G]^2$ [**Ar81**]. (This contrasts strongly with the case $\pi_X \cong \mathbb{Z}^3$, for then P must be a free module.) At the end of the chapter we mention briefly an example with $\pi_X \cong \pi_1 Kb$ and for which $\pi_2 X$ is stably free but not free.

9.3. Examples

If $(n-1, \ell) = (n-1, m)$ then $\mathbb{Z}/\ell\mathbb{Z} \rtimes_n \mathbb{Z}$ and $\mathbb{Z}/m\mathbb{Z} \rtimes_n \mathbb{Z}$ have isomorphic abelianizations. Since they have balanced presentations, every such pair of groups can be realized by an embedding, by Theorem 2.8.

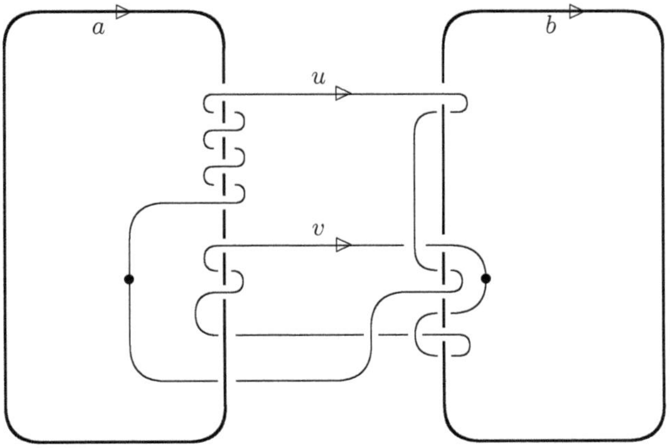

Figure 9.1 A link giving a nilpotent embedding.

The simplest non-abelian nilpotent example corresponds to the choice $\ell = 2, m = 4$ and $n = -1$. One group is $\mathbb{Z}/4\mathbb{Z} \rtimes_{-1} \mathbb{Z}$, and the other is its abelianization $\mathbb{Z} \oplus \mathbb{Z}/2\mathbb{Z}$. We shall give an explicit construction of an embedding realizing this pair of groups. Let $M = M(L)$, where L is the 4-component bipartedly trivial link depicted in Figure 9.1. If X and Y are the complementary regions for j_L then π_X and π_Y have presentations $\langle a, b \mid U = V = 1\rangle$ and $\langle u, v \mid A = B = 1\rangle$, respectively, where the words $A = u^4v^2$, $B = vuv^{-1}u^{-1}$, $U = a^4$ and $V = b^{-1}aba$, are easily read from the diagram. Since $\pi_X \cong \mathbb{Z}/4\mathbb{Z} \rtimes_{-1} \mathbb{Z}$ and $\pi_Y \cong \mathbb{Z} \oplus \mathbb{Z}/2\mathbb{Z}$, the embedding j_L is nilpotent.

It is easy to find a 4-component link $L = L_a \cup L_b \cup L_u \cup L_v$ with each 2-component sublink trivial, such that L_a and L_b represent (the conjugacy classes of) $A = [u, [u, v]]$ and $B = [v, [u, v]]$ in $F(u, v)$, respectively, while L_u and L_v have image 1 in $F(a, b)$. Arrange the link diagram so that L_u is on the left, L_v on the right, L_a at the top and L_b at the bottom. We may pass one bight of L_a which loops around L_u under a similar bight of L_b, so that U now represents $[a, b]$ in $F(a, b)$. Finally we use claspers to modify L_u and L_v so that they represent $[b, v]$ in $F(b, v)$ and $[a, u]$ in $F(a, u)$. We obtain the link of Figure 9.2.

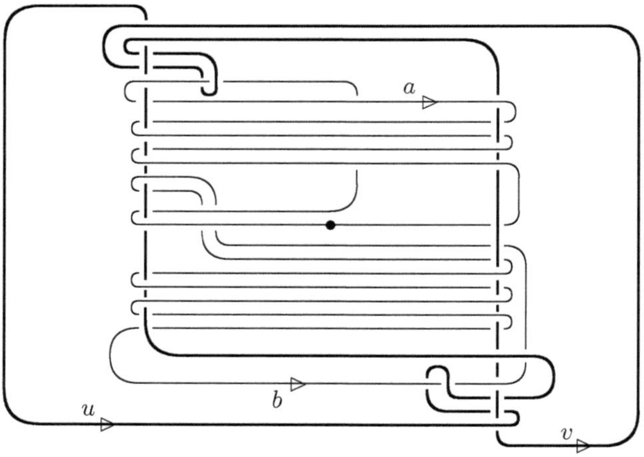

Figure 9.2 A 4-component link.

This link may be partitioned into two trivial links in three distinct ways, giving three embeddings of $M(L)$. If the two sublinks are $L_a \cup L_b$ and $L_u \cup L_v$ then $A = vu^{-1}v^{-1}u^{-1}vuv^{-1}u$, $B = vuv^{-1}u^{-1}v^{-1}uvu^{-1}$, $U = b^{-1}aba^{-1}$ and $V = 1$. Hence $\pi_X \cong \Gamma_1$ and $\pi_Y \cong \mathbb{Z}^2$.

Each of the other partitions determine abelian embeddings, with $\pi_X \cong \pi_Y \cong \mathbb{Z}^2$ and $\chi(X) = \chi(Y) = 1$. Are these two abelian embeddings equivalent?

With a little more effort, instead of passing just one bight of L_a under L_b (as above), we may interlace the loops of L_a and L_b around each of L_u and L_v so that u

and V represent $[a,[a,b]]$ and $[b,[a,b]]$, respectively, and so that each 2-component sublink of L is still trivial. If we then use claspers again we may arrange that u represents $[a,v]$ and v represents $[b,u]$, so that we obtain a 3-manifold which has one embedding with $\pi_X \cong \pi_Y \cong \Gamma_1$ and another with $\pi_X \cong \pi_Y \cong \mathbb{Z}^2$. Can we refine this construction so that the third embedding has $\pi_X \cong \Gamma_1$ and $\pi_Y \cong \mathbb{Z}^2$?

9.4. Restrained Embeddings

With our present understanding, the application of surgery in dimension 4 is limited to situations where the relevant fundamental group is in the class SA [**FT95**]. In particular, all such groups are restrained.

THEOREM 9.7. *Let $j: M \to S^4$ be a restrained embedding determined by a bipartedly trivial link. If $\beta = \beta_1(M;\mathbb{Q})$ is odd then $\chi(X) = 0$ and $\chi(Y) = 2$, and X is aspherical. If, moreover, π_X is almost coherent or elementary amenable then $\pi_X \cong \mathbb{Z}$ or $BS(1,m)$, for some $m \neq 0$, and $\beta = 1$ or 3.*

If β is even then $\chi(X) = \chi(Y) = 1$ and π_X and π_Y have balanced presentations. Moreover, if π_X and π_Y are virtually solvable then $\beta = 0$, 2, 4, 6 or 8.

PROOF. Since the embedding j derives from a bipartedly trivial link the complementary regions are each homotopy equivalent to finite 2-complexes, and j is bi-epic. Therefore $\chi(X), \chi(Y) \geqslant 0$, since π_X and π_Y are each restrained [**FMGK**, Theorem 2.5]. Hence $\chi(X)$ and $\chi(Y)$ are determined by the parity of β, since $0 \leqslant \chi(X) \leqslant \chi(Y) \leqslant 2$ and $\chi(X) \equiv \chi(Y) \equiv 1 + \beta \mod (2)$.

The asphericity of X when β is odd and π_X is restrained follows from [**FMGK**, Chapter 2.2]. We sketch the argument here. If $\chi(X) = 0$ then $H_1(X)$ is infinite, and so π_X maps onto $\mathbb{Z}$. Since π_X is finitely presentable it is then an HNN extension with finitely generated base, and since it is restrained the HNN extension is ascending. Hence $\beta_1^{(2)}(\pi) = 0$, by [**FMGK**, Lemma 2.1], and so X is aspherical, by [**FMGK**, Theorem 2.4].

If, moreover, π_X is elementary amenable or almost coherent then $\pi_X \cong \mathbb{Z}$ or $BS(1,m)$ for some $m \neq 0$, by [**FMGK**, Corollary 2.6.1]. Hence $\beta = \beta_1(X;\mathbb{Q}) + \beta_2(X;\mathbb{Q}) = 1$, if $\pi_X \not\cong BS(1,1) = \mathbb{Z}^2$, and $\beta = 3$ if $\pi_X \cong \mathbb{Z}^2$.

If β is even then $\chi(X) = 1$, and so π_X has a balanced presentation, since X is homotopy equivalent to a finite 2-complex. Similarly for π_Y.

The final assertion follows from [**Wi91**, Theorem B], which asserts that if G is a group such that G^{ab} can be generated by d elements then either G has a presentation of deficiency $\leqslant d - \frac{1}{4}(d^2-1)$, or G has an infinite quotient which is a p-torsion group, for some prime p. Since finitely generated solvable torsion groups are finite, it follows that if π_X is virtually solvable and has a balanced presentation then $\beta_1(\pi_X) \leqslant 4$, and similarly for π_Y. Hence $\beta \leqslant 8$. □

Do the conclusions hold for any bi-epic restrained embedding? (See also Lemma 7.11.)

There are examples with $\beta \leqslant 4$ or $\beta = 6$ (see Section 8.2), but we do not know of any such embeddings with $\beta = 8$. There is also a partial converse. If

$\chi(X) = 0$ and $\pi_X \cong BS(1, m)$ for some $m \neq 0$ then X is aspherical and j_{X*} is an epimorphism, by Theorem 7.10.

We conclude with an example which just fails to be nilpotent, but which illustrates the issue raised at the end of Section 9.2. This is based on the recent result of W. H. Mannan that there is a finite 2-complex C with $\pi_1 C \cong \pi_1 Kb$ and $\chi(C) = 1$ which is not homotopy equivalent to $Kb \vee S^2$, since $\pi_2 C$ is not a free $\mathbb{Z}[G]$-module [**Ma24**]. This complex corresponds to the presentation

$$\langle x, y \mid y^{-2}xy^2x^{-1} = x^{-3}y^{-1}xyx^2y^{-1}x^{-2}y = 1\rangle.$$

The 4-component link in Figure 9.3 determines an embedding of a 3-manifold with complementary regions $X_L \simeq C$ and $Y_L \simeq \mathbb{RP}^2 \vee S^2$. (This embedding is virtually abelian, but not nilpotent.)

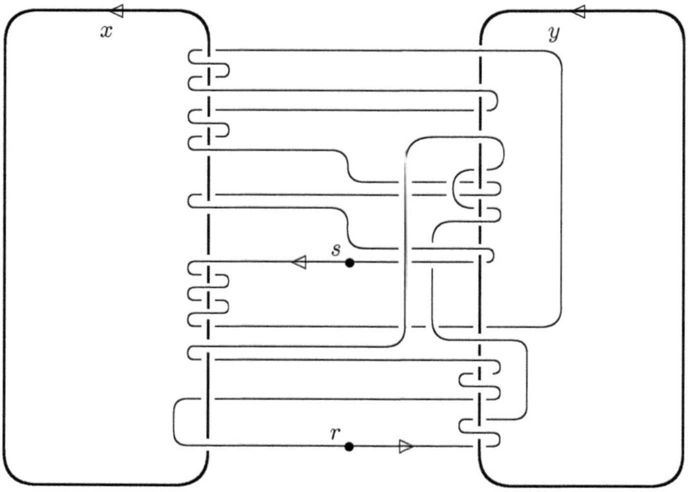

Figure 9.3 $r = y^{-2}yy^{-1}xy^2x^{-1}$, $s = x^{-3}y^{-1}xyx^2y^{-1}x^2y$.

Appendix A The Linking Pairings of Orientable Seifert Manifolds

The linking pairings of oriented 3-manifolds which are Seifert fibred over non-orientable base orbifolds were computed in Chapter 4. Here we shall consider the remaining case, when the base orbifold is also orientable. Thus the Seifert fibration is induced by a fixed-point free S^1-action on the manifold.

Bryden and Deloup have used the cohomological formulation of the linking pairing to show that every linking pairing on a finite abelian group of odd order is realized by some Seifert fibred $\mathbb{Q}$-homology sphere [**BD04**]. We shall work directly with the geometric definition, giving a new proof of this result, and shall show that there are pairings on 2-primary groups which are not realized by any orientable Seifert fibred 3-manifold at all (i.e., even if we allow non-orientable base orbifolds).

Our strategy shall be to localize at a prime p. This is a natural approach to the study of linking pairings, as they split uniquely as orthogonal sums of pairings on p-primary groups. Unfortunately, there is a loss of information in the transition from Seifert data to linking pairing, in that the linking pairing only "sees" the p-power factors of the cone point orders.

A.1 The Torsion Linking Pairing

Let $M = M(g; S)$ be a Seifert manifold with Seifert data $S = ((\alpha_1, \beta_1), \ldots, (\alpha_r, \beta_r))$, where $g \geqslant 0$, $r \geqslant 1$ and $\alpha_i > 1$ for all $i \leqslant r$. Then $H_1(M) \cong \mathbb{Z}^{2g} \oplus U$, where U has a presentation

$$\langle q_1, \ldots, q_r, h \mid \Sigma q_i = 0,\ \alpha_i q_i + \beta_i h = 0,\ \forall i \geqslant 1 \rangle.$$

The torsion subgroup τ_M is a subgroup of U.

We shall modify this presentation to obtain one with more convenient generators. Our approach involves localizing at a prime p. After reordering the Seifert data, if necessary, we may assume that α_{i+1} divides α_i in $\mathbb{Z}_{(p)}$, for all $i \geqslant 1$. (Note that $\nu = \alpha_1 \varepsilon(M)$ is then in $\mathbb{Z}_{(p)}$, while if $\varepsilon(M) = 0$ then $\frac{\alpha_1}{\alpha_2}$ is invertible in $\mathbb{Z}_{(p)}$.) Localization loses nothing, since ℓ_M is uniquely the orthogonal sum of pairings on the p-primary summands of τ_M. (We shall often write ℓ_M rather than $\mathbb{Z}_{(p)} \otimes \ell_M$, for simplicity of notation.)

Using the relation $\Sigma q_i = 0$ to eliminate the generator q_1, we see that $\mathbb{Z}_{(p)} \otimes H$ has the equivalent presentation

$$\langle q_2, \ldots, q_r, h \mid \alpha_1 \varepsilon(M) h = 0,\ \alpha_i q_i + \beta_i h = 0,\ \forall i \geqslant 2 \rangle.$$

If $r = 1$ this group is cyclic, generated by the image of h. We shall assume henceforth that $r \geqslant 2$, since all pairings on finite cyclic groups are realizable by lens spaces. Then there are integers m, n such that $m\alpha_2 + n\beta_2 = 1$, since $(\alpha_2, \beta_2) = 1$. Let $\gamma_i = \frac{\alpha_2}{\alpha_i}\beta_i$ and $q_i' = \gamma_2 q_i - \gamma_i q_2$, for all i. (Then $q_2' = 0$.) Let $s = -mh + nq_2$ and $t = \alpha_2 q_2 + \beta_2 h$. Then $h = -\alpha_2 s + nt$ and $q_2 = \beta_2 s + mt$. Since $t = 0$ in U this simplifies to

$$\langle q_3', \ldots, q_r', s \mid \alpha_1\alpha_2\varepsilon(M)s = 0,\ \alpha_i q_i' = 0,\ \forall i \geqslant 3\rangle.$$

In particular, $U = \tau_M$ if and only if $\varepsilon(M) \neq 0$, in which case $M(0; S)$ is a $\mathbb{Q}$-homology sphere.

If exactly r_p of the cone point orders α_i are divisible by p and $\varepsilon(M) = 0$ then τ_M has non-trivial p-torsion if and only if $r_p \geqslant 3$, in which case $\mathbb{Z}_{(p)} \otimes \tau_M$ is the direct sum of $r_p - 2$ cyclic submodules, while if $\varepsilon(M) \neq 0$ then τ_M has non-trivial p-torsion if and only if $r_p \geqslant 2$ and then $\mathbb{Z}_{(p)} \otimes \tau_M$ is the direct sum of $r_p - 1$ cyclic submodules. (Note however that if $r_p \leqslant 1$ and $\varepsilon(M) \neq 0$ then $\mathbb{Z}_{(p)} \otimes \tau_M \cong \mathbb{Z}_{(p)}/\alpha_1\varepsilon_S\mathbb{Z}_{(p)}$, and may be non-trivial.)

The Seifert structure gives natural 2-chains relating the 1-cycles representing the generators of U.

[For let N_i be a torus neighborhood of the ith exceptional fibre, and let B_o be a section of the restriction of the Seifert fibration to $M* = M \setminus \cup intN_i$. Let ξ_i and θ_i be simple closed curves in ∂N_i which represent q_i and h, respectively. Then $\partial B_o = \Sigma\xi_i$, and there are singular 2-chains D_i in N_i such that $\partial D_i = \alpha_i\xi_i + \beta_i\theta_i$, since $\alpha_i q_i + \beta_i h = 0$ in $H_1(N_i)$.]

We may choose disjoint annuli A_i in M^* with $\partial A_i = \theta_2 - \theta_i$, for $i \neq 2$. For convenience in our formulae, we shall also let $A_2 = 0$. Then $C_i = \beta_2 D_i - \beta_i D_2 + \beta_2\beta_i A_i$ is a singular 2-chain with $\partial C_i = \alpha_i\beta_2\xi_i - \alpha_2\beta_i\xi_2$.

Let $\xi_i' = \gamma_2\xi_i - \gamma_i\xi_2$, for $i \geqslant 3$, $\sigma = -m\theta_2 + n\xi_2$ and

$$U = \alpha_1 B_o + \alpha_1\varepsilon_S n D_2 - \Sigma\frac{\alpha_1}{\alpha_i}(D_i + \beta_i A_i).$$

Then ξ_i' is a singular 1-chain representing q_i' and $\partial C_i = \alpha_i\xi_i'$, for all $i \geqslant 3$, σ is a singular 1-chain representing s and U is a singular 2-chain with $\partial U = \alpha_1\alpha_2\varepsilon_S\sigma$.

We may assume that $\xi_i \bullet \theta_i = 1$ in ∂N_i. In order to calculate intersections and self-intersections of the 1-cycles ξ_i with the 2-chains C_i in M, we may push each ξ_i off N_i. Then ξ_i and D_j are disjoint, for all i, j, while $\xi_2 \bullet A_i = 1$, $\xi_i \bullet A_i = -1$ and $\xi_j \bullet A_i = 0$, if $i, j \neq 2$ and $j \neq i$. Similarly, we may assume that θ_2 is disjoint from the discs D_j (for all j) and the annuli A_k (for all $k \neq 2$). Since B_o is oriented so that $\partial B_o = \Sigma\xi_i$, we must have $\theta_2 \bullet B_o = -1$. Hence

$$\xi_i' \bullet C_i = -\beta_2\beta_i(\gamma_2 + \gamma_i),$$

$$\xi_i' \bullet C_j = -\beta_2\beta_j\gamma_i,$$

and

$$\xi_i' \bullet U = \alpha_1\varepsilon_S\gamma_i$$

for all $i, j \geqslant 3$ with $j \neq i$, while

$$\sigma \bullet U = -\frac{\alpha_1}{\alpha_2} - n\alpha_1\varepsilon_S$$

and

$$\sigma \bullet C_i = n\beta_2\beta_i.$$

Then

$$\ell_M(q_i', q_i') = [-\beta_2\beta_i \frac{\alpha_i\beta_2 + \alpha_2\beta_i}{\alpha_i^2}] \in \mathbb{Q}/\mathbb{Z}$$

and

$$\ell_M(q_i', q_j') = [-\beta_2\beta_i\beta_j \frac{\alpha_2}{\alpha_i\alpha_j}] \in \mathbb{Q}/\mathbb{Z}.$$

If $\varepsilon(M) \neq 0$ then the above calculations of $\xi_i' \bullet U$ and $\sigma \bullet C_i$ each give

$$\ell_M(s, q_i') = [\frac{\beta_i}{\alpha_i}] \in \mathbb{Q}/\mathbb{Z}$$

and

$$\ell_M(s, s) = [-\frac{\alpha_1 + n\alpha_1\alpha_2\varepsilon(M)}{\alpha_1\alpha_2^2\varepsilon(M)}] \in \mathbb{Q}/\mathbb{Z}.$$

In particular, the linking pairings depend only on S and not on g. (We could arrange that the deminators are powers of p, after further rescaling the basis elements. However, that would tend to obscure the dependence on the Seifert data.)

Let S and S' be two systems of Seifert data, with concatenation S'', and let $M'' = M\#_f M' = M(0; S'')$ be the fibre-sum of $M = M(0; S)$ and $M' = M(0; S')$. Then $\varepsilon(M'') = \varepsilon(M) + \varepsilon(M')$. The next result is clear.

LEMMA A.1. *Let $M = M(g; S)$ and $M' = M(g'; S')$ be Seifert manifolds such that all the cone point orders of S' are relatively prime to all the cone point orders of S and $\varepsilon(M') = \varepsilon(M) = 0$, and let $M'' = M(g + g'; S'')$. Then $\varepsilon(M'') = 0$ and $\ell_{M''} = \ell_M \perp \ell_{M'}$.* □

Thus, if every p-primary summand of a linking pairing ℓ can be realized by some $M(0; S)$ with all cone point orders powers of p and $\varepsilon(M) = 0$, then ℓ can also be realized by a Seifert manifold. If one of the hypotheses fails, it is not clear how the linking pairings of M, M' and M'' are related. In order to realize pairings by Seifert manifolds with $\varepsilon(M) \neq 0$ we shall need another approach.

A.2 The Homogeneous Case: p Odd

In this section we shall show that when p is odd and the p-primary component of τ_M is homogeneous the structure of the p-primary component of ℓ_M may be read off the Seifert data. Our results shall be extended to the inhomogeneous cases in later sections. Let $u_i = \frac{\alpha_i}{p^k}$, for $1 \leqslant k \leqslant r_p$, and $v = \alpha_1\varepsilon(M)$.

LEMMA A.2. *Let $M = M(g; S)$ be a Seifert manifold and p a prime. Then $\mathbb{Z}_{(p)} \otimes \tau_M$ is homogeneous of exponent p^k if and only if either*

(1) *$\varepsilon(M) = 0$ and u_i is invertible in $\mathbb{Z}_{(p)}$, for $1 \leqslant k \leqslant r_p$; or*
(2) *$p^{-k}\alpha_1\varepsilon(M)$ and u_i are invertible in $\mathbb{Z}_{(p)}$, for $2 \leqslant k \leqslant r_p$; or*
(3) *$r_p \leqslant 2$ and $p^{-k}\alpha_1\alpha_2\varepsilon(M)$ is invertible in $\mathbb{Z}_{(p)}$.*

PROOF. This follows immediately from the calculations in the first section, with the following observations. If $\varepsilon(M) = 0$ then α_1 and α_2 must have the same p-adic valuation. If $r_p \geqslant 2$ then $u_2 = p^{-k}\alpha_2$ and $v = \alpha_1\varepsilon(M)$ are in $\mathbb{Z}_{(p)}$. Hence if $u_2 v = p^{-k}\alpha_1\alpha_2\varepsilon(M)$ is invertible in $\mathbb{Z}_{(p)}$ then u_2 and v are also invertible in $\mathbb{Z}_{(p)}$. □

Note that if $\mathbb{Z}_{(p)} \otimes \tau_M$ is homogeneous of exponent p^k and $\varepsilon(M) \neq 0$ then u_1 may be divisible by p.

THEOREM A.3. *Let $M = M(g; S)$ be a Seifert manifold and p an odd prime such that $\mathbb{Z}_{(p)} \otimes \tau_M$ is homogeneous of exponent p^k. Then*

(1) *if $\varepsilon(M) = 0$ then $d(\ell_M) = [(-1)^{r_p-1}\frac{\alpha_1}{\alpha_2}(\Pi_{i=1}^{r_p}\beta_i)(\Pi_{j=3}^{r_p}u_j)]$;*
(2) *if $\varepsilon(M) \neq 0$ then $d(\ell_M) = [(-1)^{r_p-1}\frac{\alpha_1}{\alpha_2}(\Pi_{i=1}^{r_p}\beta_i)(\Pi_{j=2}^{r_p}u_j)v]$.*

PROOF. Let $L \in \mathrm{GL}(\rho, \mathbb{Z}/p^k\mathbb{Z})$ be the matrix with (i,j) entry $p^k\ell(e_i, e_j)$, where $e_1, \dots, e_\rho$ is some basis for $\mathbb{Z}_{(p)} \otimes \tau_M \cong (\mathbb{Z}/p^k\mathbb{Z})^\rho$.

Suppose first that $\varepsilon(M) = 0$. Then $\mathbb{Z}_{(p)} \otimes \tau_M \cong (\mathbb{Z}/p^k\mathbb{Z})^{r_p-2}$, with basis $e_i = q'_{i+2}$, for $1 \leqslant i \leqslant r_p - 2$, and $\frac{\alpha_1}{\alpha_2} = \frac{u_1}{u_2}$ is invertible in $\mathbb{Z}_{(p)}$. We apply row operations

$$row_i \mapsto row_i - \frac{\alpha_3\beta_{i+2}}{\alpha_{i+2}\beta_3} row_1$$

for $2 \leqslant i \leqslant r_p - 2$ and then

$$row_1 \mapsto row_1 - \frac{\alpha_2\beta_3}{\alpha_3\beta_2}\Sigma_{i=2}^{r_p-2} row_i$$

to L. This gives a lower triangular matrix with diagonal

$$[-\beta_2\frac{\beta_3}{u_3}(\beta_2 + u_2\Sigma_{i=3}^{r_p}\frac{\beta_i}{u_i}), -\beta_2^2\frac{\beta_4}{u_4}, \dots, -\beta_2^2\frac{\beta_{r_p-2}}{u_{r_p-2}}].$$

Therefore,

$$\det(L) = (-1)^{r_p}\beta_2^{2r_p-3}(\beta_2 + u_2\Sigma_{i=3}^{r_p}\frac{\beta_i}{u_i})\Pi_{i=3}^{r_p}\frac{\beta_j}{u_j},$$

and so

$$d(\ell_M) = [(-1)^{r_p-1}\frac{\alpha_2}{\alpha_1}(\Pi_{i=1}^{r_p}\beta_i)\Pi_{j=3}^{r_p}u_j)].$$

A similar argument applies if $\varepsilon(M) \neq 0$. In this case u_2 and $v = \alpha_1\varepsilon(M)$ are also invertible in $\mathbb{Z}_{(p)}$, and $\mathbb{Z}_{(p)} \otimes \tau_M \cong (\mathbb{Z}/p^k\mathbb{Z})^{r_p-1}$, with basis $e_i = q'_{i+2}$, for $1 \leqslant i \leqslant r_p - 2$, and $e_{r_p-1} = s$. If we perform the same row operations on rows 2 to $r_p - 2$, and then the column operation $col_1 \mapsto col_1 + \Sigma_{i=2}^{r_p-2} col_i$ we obtain a bordered matrix

$$\begin{pmatrix} -\beta_2\frac{\beta_3}{u_3}(\beta_2 + u_2\Sigma^*) & 0 & \dots & 0 & \frac{\beta_3}{u_3} \\ 0 & -\beta_2^2\frac{\beta_4}{u_4} & \dots & 0 & 0 \\ \vdots & 0 & \ddots & \vdots & \vdots \\ 0 & \dots & 0 & -\beta_2^2\frac{\beta_{r_p}}{u_{r_p}} & 0 \\ \Sigma^* & \frac{\beta_4}{u_4} & \dots & \frac{\beta_{r_p}}{u_{r_p}} & d^* \end{pmatrix}$$

where $\Sigma^* = \Sigma_{i=3}^{r_p} \frac{\beta_i}{u_i}$ and $d^* = -\frac{u_1+u_2 v}{u_2^2 v}$. Hence

$$\det(L) = -(\beta_2 \frac{\beta_3}{u_3}(\beta_2 + u_2\Sigma^*)d^* + \frac{\beta_3}{u_3}\Sigma^*)(-1)^{r_p-3}\beta_2^{2(r_p-3)}\Pi_{i=4}^{r_p}\frac{\beta_i}{u_i}$$

$$= (-\beta_d d^*(\beta_2 + u_2\Sigma^*) + \Sigma^*)(-1)^{r_p-2}\beta_2^{2(r_p-3)}\Pi_{i=3}^{r_p}\frac{\beta_i}{u_i}.$$

Now

$$(-\beta_2 d^*(\beta_2 + u_2\Sigma^*) + \Sigma^*) = -(\beta_2(u_1 + u_2 v)(-\beta_2 + u_2\Sigma^*) - u_2^2\nu\Sigma^*)/u_2^2\nu$$

$$= (\beta_2(u_1\beta_2 + u_1 u_2\Sigma^* + n\beta_2 u_2 v) + (n\beta_2 - 1)u_2^2 v\Sigma^*)/u_2^2 v$$

$$\equiv \beta_2(u_1\beta_2 + u_1 u_2\Sigma^* + u_2 v) \equiv -\beta_1\beta_2 u_2 \quad mod\ (p),$$

since $n\beta_2 \equiv 1\ mod\ (p)$ and $v = -\beta_1 - \frac{\beta_1 u_1}{u_2} - u_1\Sigma^*$. Therefore

$$\det(L) \equiv (-1)^{r_p-1}\beta_2^{2(r_p-3)}(\Pi_{i=1}^{r_p}\beta_i)/\Pi_{j=2}^{r_p}u_j v \quad mod\ (p),$$

and so we now have

$$[d(\ell_M) = [(-1)^{r_p-1}(\Pi_{1\leqslant i\leqslant r_p}\beta_i)(\Pi_{2\leqslant j\leqslant r_p}u_j)v].$$

□

When all the cone point orders have the same p-adic valuation (i.e., u_1 and u_2 are also invertible in $\mathbb{Z}_{(p)}$) then these formulae for $d(\ell_M)$ are invariant under permutation of the indices. For if $\varepsilon(M) = 0$ then $[\frac{\alpha_1}{\alpha_2}] = [u_1 u_2]$ in $\mathbb{F}_p^\times/(\mathbb{F}_p^\times)^2$, while if $\varepsilon(M) \neq 0$ then $\nu = u_1 p^k \varepsilon(M)$ (and $p^k\varepsilon(M)$ is also invertible).

A linking pairing ℓ on a free $\mathbb{Z}/p^k\mathbb{Z}$-module N is hyperbolic if and only if $\rho = rk(\ell)$ is even and $d(\ell) = (-1)^{\frac{\rho}{2}}$. Thus $\mathbb{Z}_{(p)} \otimes \ell_M$ is hyperbolic if and only if either $\varepsilon(M) = 0$, $r_p = \rho + 2$ is even and $[\frac{\alpha_1}{\alpha_2}(\Pi_{1\leqslant i\leqslant r_p}\beta_i)(\Pi_{3\leqslant j\leqslant r_p}u_j)] = [(-1)^{\frac{r_p}{2}}]$, or $\varepsilon(M) \neq 0$, $r_p = \rho + 1$ is odd and $[(\Pi_{1\leqslant i\leqslant r_p}\beta_i)(\Pi_{2\leqslant j\leqslant r_p}u_j)v] = [(-1)^{\frac{r_p-1}{2}}]$. In particular, if S is skew-symmetric and all cone point orders divisible by p are divisible by the same power of p then $\alpha_1 = \alpha_2$ and $\Pi_{1\leqslant i\leqslant r_p}\beta_i = (-1)^{\frac{r_p}{2}}\Pi_{1\leqslant i\leqslant r_p}|\beta_i|$. Since $\Pi_{1\leqslant i\leqslant r_p}|\beta_i|$ and $\Pi_{3\leqslant j\leqslant r_p}u_j$ are squares, we see that $\mathbb{Z}_{(p)} \otimes \ell_M$ is hyperbolic.

A.3 Realization of Pairings on Groups of Odd Order

In this section we shall show that every linking pairing on a finite group of odd order may be realized by a Seifert manifold.

Suppose first that we localize ℓ_M at a prime p. Let p^k be the exponent of $\mathbb{Z}_{(p)} \otimes \tau_M$, and let L be the matrix with entries $p^k\ell(q_i', q_j')$. (If $\varepsilon(M) \neq 0$ we need also a row and column corresponding to the generator s, which has the maximal order p^k.) Then

$$L = \begin{pmatrix} D_1 & p^{\kappa_2}B_2 & \dots & p^{\kappa_t}B_t \\ p^{\kappa_2}B_2^{tr} & p^{\kappa_2}D_2 & \dots & \vdots \\ \vdots & \dots & \ddots & \vdots \\ p^{\kappa_t}B_t^{tr} & \dots & \dots & p^{\kappa_t}D_t \end{pmatrix}$$

where D_i is a $\rho_i \times \rho_i$ block with $\det(D_i) \not\equiv 0\ mod\ (p)$, for $1 \leqslant i \leqslant t$, and $0 < \kappa_2 < \dots < \kappa_t < k$. We may partition L more coarsely as $L = \left(\begin{smallmatrix} A & B \\ B^{tr} & D \end{smallmatrix}\right)$, where $A = D_1$

and $B = [B_2 \dots B_t]$. Let $Q = \begin{pmatrix} I-d_1^{-1} & p^{\kappa_2}B \\ 0 & I \end{pmatrix}$ and $D' = D - p^{\kappa_2}B^{tr}A^{-1}B$. Then $Q^{t}rLQ = \begin{pmatrix} A & 0 \\ 0 & p^{\kappa_2}D' \end{pmatrix}$. Block-diagonalizing L in this fashion does not change the residues *mod* (p) of the diagonal blocks D_i or decrease the divisibility of the off-diagonal blocks. These matrix manipulations correspond to replacing the generators q_i' with $i > \rho_1$ by $\widetilde{q}_i = q_i' - p^{\kappa_2}\Sigma_{j=1}^{\rho_1}[B^{tr}A]_{ji}q_j'$.

We may iterate this process, and we find that $\mathbb{Z}_{(p)} \otimes \ell_M$ is an orthogonal sum of pairings on homogeneous groups $(\mathbb{Z}/p^k\mathbb{Z})^{\rho_1}$, $(\mathbb{Z}/p^{k-\kappa_2}\mathbb{Z})^{\rho_2}, \dots, (\mathbb{Z}/p^{k-\kappa_t}\mathbb{Z})^{\rho_t}$. If $\varepsilon(M) = 0$ or if $\alpha_1\varepsilon(M)$ is invertible in $\mathbb{Z}_{(p)}$ then the determinantal invariants of the first summand (with the maximal exponent p^k) may be computed from block A as in Theorem A3, while we may read off the determinantal invariants of the other summands from the corresponding diagonal elements of the original matrix L. (We shall not need to consider the possibility that p divides the numerator of ε_S in justifying our constructions below.)

It follows easily that if the Seifert data S is skew-symmetric then each of the homogeneous orthogonal summands of $\mathbb{Z}_{(p)} \otimes \ell_M$ is hyperbolic, and hence $\mathbb{Z}_{(p)} \otimes \ell_M$ is hyperbolic when p is odd. If, moreover, all cone points are odd then ℓ_M is hyperbolic. (This is to be expected, as Seifert manifolds with such Seifert data embed in S^4, but a purely algebraic argument is preferable.)

With these reductions in mind, we may now construct Seifert manifolds realizing given pairings.

THEOREM A.4. *Let p be an odd prime, and let ℓ be a linking pairing on a finite abelian p-group. Then there is a Seifert manifold $M = M(0; S)$ such that all the cone point orders α_i are powers of p, $\varepsilon(M) = 0$ and $\ell_M \cong \ell$.*

PROOF. The pairing ℓ_M is the orthogonal sum $\perp_{j=1}^{t} \ell_j$, where ℓ_j is a pairing on $(\mathbb{Z}/p^{k_j}\mathbb{Z})^{\rho_j}$, with $\rho_j > 0$ for $1 \leqslant j \leqslant t$ and $0 < k_k < k_{j-1}$ for $2 \leqslant j \leqslant t$. Let $d(\ell_j) = [w_j]$ for $1 \leqslant j \leqslant t$, and let $k = k_1$.

If $p \geqslant 5$ we let $\alpha_i = p^k$ for $1 \leqslant i \leqslant m_1 = \rho_1 + 2$, $\alpha_i = p^{k_2}$ for $m_1 < i \leqslant m_2 = m_1 + \rho_2, \dots$, and $\alpha_i = p^{k_t}$ for $m_{t-1} < i \leqslant r = (\Sigma\rho_j) + 2$. For each $1 \leqslant j \leqslant t$ we let $\beta_i = 1$ for $m_j < i < m_{j+1}$ and $\beta_{m_{j+1}} = w_j$. We must then choose β_i for $1 \leqslant i \leqslant m_1$ so that $[\Pi_{i=1}^{m_1}\beta_i] = [w_1]$ and $\Sigma_{i=1}^{m_1}\beta_i = -\Sigma_{j=m_1+1}^{r}p^k\frac{\beta_j}{\alpha_j}$. It is in fact sufficient to solve the equations $[\Pi_{i=1}^{m_1}\beta_i] = [w_1]$ and $\Sigma_{i=1}^{m_1}\beta_i = 0$ with all $\beta_i \in \mathbb{Z}/p^k\mathbb{Z}^\times$, for subtracting $\Sigma_{j=m_1+1}^{r}p^k\frac{\beta_j}{\alpha_j}$ from β_1 will not change its residue *mod* (p).

If m_1 is odd the equation $\Sigma\beta_i = 0$ always has solutions with all $\beta_i \in \mathbb{Z}/p^k\mathbb{Z}^\times$. If ξ is a non-square in $\mathbb{Z}/p^k\mathbb{Z}^\times$ setting $\beta_i' = \xi\beta_i$ for all i gives another solution, and $[\Pi\beta_i'] = [\xi][\Pi\beta_i]$. (If $p \equiv 3$ *mod* (4) we may take $\xi = -1$, which corresponds to a change of orientation of the 3-manifold.)

If $m_1 = 4t$ and $w \not\equiv 1$ *mod* (p) then there is an integer x such that $2x \equiv w - 1$ *mod* (p). The images of x and $w - 1 - x$ are invertible in $\mathbb{Z}/p^k\mathbb{Z}$. Let $\beta_1 = 1$, $\beta_2 = -w$, $\beta_3 = x$ and $\beta_4 = w - 1 - x$, and $\beta_{2i+1} = 1$ and $\beta_{2k+2} = -1$ for $2 \leqslant i \leqslant 2t$. Then $\Sigma\beta_i = 0$, $\beta_4 \equiv \beta_3$ *mod* (p) and $[(-1)^{r-1}\Pi\beta_i] = [w]$.

If $\rho = 4t + 2$ and $w \not\equiv 1$ *mod* (p), let y be an integer such that $4y \equiv w + 1$ *mod* (p). Let $\beta_1 = 1$, $\beta_2 = -w$, $\beta_3 = \beta_4 = \beta_5 = y$ and $\beta_6 = w - 1 - 3y$, and $\beta_{2i+1} = 1$ and $\beta_{2i+2} = -1$ for $3 \leqslant i < 2t$. Then $\Sigma\beta_i = 0$, $\beta_6 \equiv \beta_3$ *mod* (p) and $[(-1)^{r-1}\Pi\beta_i] = [w]$.

These choices work equally well for all $p \geqslant 3$, if $[w] \neq 1$. If ρ is even, $w \equiv 1$ *mod* (p) and $p > 3$ there is an integer n such that $n^2 \neq 0$ or 1 *mod* (p), and we solve as before, after replacing w by $\widehat{w} = n^2 w$.

However, if $p = 3$ and $[w] = 1$ we must vary our choices. If $\rho = 4t - 2$ with $t > 1$, let $\beta_1 = \beta_2 = \beta_3 = \beta_4 = 1$, $\beta_5 = \beta_6 = -2$ and $\beta_{2i+1} = 1$ and $\beta_{2i+2} = -1$ for $3 \leqslant i < 2t$. If $\rho = 4t$ let $\beta_{2i-1} = 1$ and $\beta_{2i} = -1$ for $1 \leqslant i \leqslant 2t + 1$. In the remaining case (when $\rho = 2$) we find that if $\Sigma_{1 \leqslant i \leqslant 4}\beta_i = 0$ then $[-\Pi\beta_i] = [-1]$. In this case we must use instead $S = ((3^{k+1}, 1), (3^{k+1}, 5), (3^k, -1), (3^k, -1))$ to realize the pairing with $[w] = [1]$. □

The manifolds with Seifert data as above are $\mathbb{H}^2 \times \mathbb{E}^1$-manifolds, except when $\rho = 1$ and $p = 3$, in which case they are the flat manifold G_3 (with its two possible orientations).

It follows immediately from Theorem A4 and Lemma A1 that every linking pairing on a finite abelian group of odd order is realized by some Seifert manifold $M(0; S)$ with $\varepsilon(M) = 0$.

All such pairings may also be realized by Seifert manifolds which are $\mathbb{Q}$-homology spheres. However, we must be careful to ensure that the *numerator* of $\varepsilon(M)$ does not provide unexpected torsion.

THEOREM A.5. *Let ℓ be a linking pairing on a finite abelian group A of odd order. Then there is a Seifert manifold $M = M(0; S)$ such that $\varepsilon(M) \neq 0$ and $\ell_M \cong \ell$.*

PROOF. Let P be the finite set of primes for which the p-primary summand of A is non-trivial, and let $\ell = \perp_{p \in P} \ell^{(p)}$ be the primary decomposition of ℓ. We shall define Seifert data $S(p)$ for each $p \in P$ as follows. Suppose that $\ell^{(p)}$ is a pairing on $\oplus_{j=1}^{t(p)} (\mathbb{Z}/p^{k_j}\mathbb{Z})^{\rho_j}$, with $\rho_j > 0$ for $1 \leqslant j \leqslant t(p)$ and $0 < k_j < k_{j-1}$ for $2 \leqslant j \leqslant t(p)$. Then $\ell^{(p)} \cong \perp \ell_{b_i/a_i}$, where the a_i are powers of p, for $1 \leqslant i \leqslant \rho(p) = \Sigma_{j=1}^{t(p)} \rho_j$ and such that $a_i \geqslant a_{i+1}$ for $i < \rho(p)$. Let $S(p) = ((\alpha_1^{(p)}, \beta_1^{(p)}), \ldots, (\alpha_{t(p)}^{(p)}, \beta_{t(p)}^{(p)}))$, where $(\alpha_i^{(p)} = a_i$, for all i, $\beta_1^{(p)} = (-1)^{\rho_1 + 1} b_1$ and $\beta_i^{(p)} = b_i$, for $2 \leqslant i \leqslant \rho(p)$. Finally let $\widetilde{\alpha} = e\Pi_{p \in P} p$, where e is the exponent of A, and let $\widetilde{\beta} = -1 - \widetilde{\alpha}\Sigma_{i=1}^{(t(p)} \frac{\beta_i^{(p)}}{\alpha_i^{(p)}}$.

Let S be the concatenation of $(\widetilde{\alpha}, \widetilde{\beta})$ and the sets $S(p)$ for $p \in P$, and let $M = M(0; S)$. Then $\varepsilon(M) = \frac{1}{\widetilde{\alpha}}$. For each $p \in P$ there are $\rho(p) + 1$ cone points with order divisible by p, and $\mathbb{Z}_{(p)} \otimes \tau_M \cong \oplus_{j=1}^{t(p)} \mathbb{Z}/p^{k_j}\mathbb{Z})^{\rho_j}$. Since $\widetilde{\beta} \equiv -1$ *mod* (p), for all $p \in P$, the determinantal invariant of the component of $\mathbb{Z}_{(p)} \otimes \ell_M$ of maximal exponent $a_1 = p^{k_1}$ is $[\Pi_{i=1}^{\rho_1} b_i]$. Therefore, $\mathbb{Z}_{(p)} \otimes \ell_M \cong \ell^{(p)}$, for each $p \in P$, and so $\ell_M \cong \ell$. □

The manifolds with Seifert data as above are $\widetilde{\mathbb{SL}}$-manifolds, except when $A \cong \mathbb{Z}/p^k\mathbb{Z}$, and so there are just two cone points, in which case they are lens spaces ($\mathbb{S}^3$-manifolds).

In the homogeneous p-primary case we may arrange that all cone points have order p^k, except when $p = 3$, $\rho = 2$ and $d(\ell) = [1]$. This case is realized by $M(0; (3^{k+1}, 7), (3^k, -1), (3^k, -1))$.

A.4 Realization of Homogeneous 2-Primary Pairings

The situation is more complicated when $p = 2$. A linking pairing ℓ on $(\mathbb{Z}/2^k\mathbb{Z})^\rho$ is determined by its rank ρ and certain invariants $\sigma_j(\ell) \in \mathbb{Z}/8\mathbb{Z} \cup \{\infty\}$, for $\rho - 2 \leqslant j \leqslant \rho$. (See §3 of [**KK80**], and [**De05**].) We shall not calculate these invariants here. Instead, we shall take advantage of the particular form of the pairings given in the first section.

If a linking pairing ℓ on $(\mathbb{Z}/2^k\mathbb{Z})^\rho$ is even then ρ is also even. When $k = 1$ all even pairings are hyperbolic. If $k > 1$ then ℓ is determined by the image of the matrix L in $GL(\rho, \mathbb{Z}/4\mathbb{Z})$, and is either hyperbolic (and is the orthogonal sum of $\frac{\rho}{2}$ copies of the pairing E_0^k) or is the orthogonal direct sum of a hyperbolic pairing of rank $\rho - 2$ with the pairing E_1^k [**KK80, Wa64**].

We shall say that an element $\frac{p}{q}$ is *even* or *odd* if p is even or odd, respectively. Thus $\frac{p}{q}$ is odd if and only if it is invertible in $\mathbb{Z}_{(2)}$.

The following result complements the criterion for homogeneity given in Lemma A2.

LEMMA A.6. *Let $M = M(g; S)$ be a Seifert manifold and assume that the Seifert data are ordered so that α_{i+1} divides α_i in $\mathbb{Z}_{(2)}$. Then*

(1) *$\ell = \mathbb{Z}_{(2)} \otimes \ell_M$ is even if and only if $\frac{\alpha_1}{\alpha_i}$ is odd for $1 \leqslant i \leqslant r_2$ and either $\varepsilon(M) = 0$ or $\alpha_1\varepsilon(M)$ is odd;*

(2) *if $\frac{\alpha_1}{\alpha_i}$ is odd for $1 \leqslant i \leqslant r_2$ then $\alpha_1\varepsilon(M) \equiv r_2$ mod (2).*

PROOF. If $\ell = \mathbb{Z}_{(2)} \otimes \ell_M$ is even then $\beta_2 + \frac{\alpha_2}{\alpha_i}\beta_i$ is even for all $3 \leqslant i \leqslant r_2$. Hence $\frac{\alpha_2}{\alpha_i}$ is odd, since each β_i is odd. If moreover $\varepsilon(M) = 0$ then $\frac{\alpha_1}{\alpha_2}$ is odd. If $\varepsilon(M) \neq 0$ then $\frac{\alpha_1}{\alpha_2} + n\alpha_1\varepsilon(M)$ is even. Hence $\frac{\alpha_1}{\alpha_2}$ is again odd, and so $\alpha_1\varepsilon(M)$ is also odd. In each case the converse is clear.

The second assertion holds since β_i is odd for $1 \leqslant i \leqslant r_2$ and $\frac{\alpha_1}{\alpha_i}$ is even for all $i > r_2$. □

We shall suppose for the remainder of this section that $\mathbb{Z}_{(2)} \otimes \tau_M$ is homogeneous of exponent $2^k > 1$.

Suppose first that $\mathbb{Z}_{(2)} \otimes \ell_M$ is even. Then it is homogeneous and of even rank $\rho = 2s$. The diagonal entries of L are all even and the off-diagonal entries are all odd. If $k = 1$ then $\mathbb{Z}_{(2)} \otimes \ell_M$ is hyperbolic, so we may assume that $k > 1$ in the next theorem.

THEOREM A.7. *Let $M = M(g; S)$ be a Seifert manifold such that the even cone point orders α_i all have the same 2-adic valuation $k > 1$. Assume that either $\varepsilon(M) = 0$ or $\alpha_1\varepsilon(M)$ is odd. Let t be the number of diagonal entries of L which* *are divisible by* 4*. Then whether* $\mathbb{Z}_{(2)} \otimes \ell_M$ *is hyperbolic or not depends only on the images of* t *and* ρ *in* $\mathbb{Z}/4\mathbb{Z}$*.*

PROOF. The linking pairing is even, by Lemma 6, and so ρ is even. We may reorder the basis of τ_M so that $L_{ii} \equiv 0$ *mod* (4), for all $i \leqslant t$ and $L_{ii} \equiv 2$ *mod* (4) for $t < i \leqslant \rho$. Let $t = 4a + x$ and $\rho - t = 4b + y$, where $0 \leqslant x, y \leqslant 3$. Then $x+y = \rho-4(a+b)$ is even, and there are eight relevant pairs (x, y). We may partition L as $L = \left(\begin{smallmatrix} E & F \\ F^{tr} & G \end{smallmatrix}\right)$, where $E = \left(\begin{smallmatrix} L_{11} & L_{12} \\ L+21 & L_{22} \end{smallmatrix}\right)$ is invertible, G is a $(\rho - 2) \times (\rho - 2)$-submatrix and F is a $2 \times (\rho - 2)$-submatrix. If we use $J = \begin{pmatrix} I_2 & -E^{-1}F \\ 0 & I_{\rho-2} \end{pmatrix}$ to obtain $J^{tr}LJ = \left(\begin{smallmatrix} E & 0 \\ 0 & G' \end{smallmatrix}\right)$, then $G' = G - F^{tr}E^{-1}F$. The entries of F are all odd and so the entries of $F^{tr}E^{-1}F$ are all congruent to 2 *mod* (4). Therefore $G' \equiv G$ *mod* (2), and so the off-diagonal entries of G' are still odd, but the residues *mod* (4) of the diagonal entries are changed. An application of this process to G' then restores the residue classes of the diagonal entries of the corresponding $(\rho - 4) \times (\rho - 4)$-submatrix. Iterating this process, we find that $\ell_M \cong (a + b)(E_0^k \perp E_1^k) \perp \ell'$, where ℓ' has rank $x + y$ and the off-diagonal entries for ℓ' are odd. We also find that $\ell' \cong (x + y)E_0^k$, unless $\{x, y\} = \{1, 3\}$ or $\{0, 2\}$, in which case $\ell' \cong E_0^k \perp E_1^k$ or E_1^k, respectively. Since E_0^k is hyperbolic, $2E_1^k \cong 2E_0^k$ [**Wa64**] and E_1^k is not hyperbolic, it follows that ℓ is hyperbolic if and only if either $a + b$ is even and $\{x, y\} \neq \{1, 3\}$ or $\{0, 2\}$, or if $a + b$ is odd and $\{x, y\} = \{1, 3\}$ or $\{0, 2\}$. □

Assume that $\mathbb{Z}_{(2)} \otimes \ell_M$ is even, and that the indexing of S is as in Lemma A6, and let $\psi_i = \frac{\alpha_i}{\alpha_2}$, for $i \leqslant r_2$. It follows immediately from Theorem A7 and the calculations in the first section that

$$t = |\{3 \leqslant i \leqslant r_2 \mid \beta_i + \psi_i\beta_2 \equiv 0 \ mod\ (4)\}| + \delta,$$

where $\delta = 1$ if $\varepsilon(M) \neq 0$ and $\beta_2 + \alpha_2\varepsilon(M) \equiv 0$ *mod* (4), and $\delta = 0$ otherwise.

For example, if $S = ((8, 1), (8, -1), (8, 3), (8, -3))$ then $\rho = 2$ and $t = 0$, so $a + b = 0$ and $x + y = 2$, and $\ell \cong E_1^3$. Thus, skew symmetry of the Seifert data does not imply that $\mathbb{Z}_{(2)} \otimes \ell_M$ is hyperbolic.

Lemmas A2 and A6 imply that if $\mathbb{Z}_{(2)} \otimes \tau_M$ is homogeneous of exponent 2^k then $\mathbb{Z}_{(2)} \otimes \ell_M$ is odd if and only if either

(1) $\varepsilon(M) = 0$, $2^{-k}\alpha_i$ is even for $i = 1$ and 2, and is odd for $2 < i \leqslant r_2$; or

(2) $\alpha_1\varepsilon(M)$ and $2^{-k}\alpha_i$ are odd for $1 < i \leqslant r_2$, and either $2^{-k}\alpha_1$ or r_2 is even.

Odd forms on homogeneous 2-groups can be diagonalized. In the present situation, this follows easily from the next lemma.

LEMMA A.8. *Let* ℓ *be an odd linking pairing on* $N = (\mathbb{Z}/2^k\mathbb{Z})^2$*. Then* ℓ *is diagonalizable.*

PROOF. Let e, f be the standard basis for N. Since ℓ is odd we may assume that $\ell(e, e) = [2^{-k}a]$, where a is odd. Let $\ell(e, f) = [2^{-k}b]$ and $\ell(f, f) = [2^{-k}d]$. (Then b is even and d is odd, or vice versa, by non-singularity of the pairing.) Let $f' = -a^{-1}be + f$. Then $\ell(e, f') = 0$ and $\ell(f', f') = [2^{-k}d']$, where $d' \equiv d - a^{-1}b^2$ *mod* (2^k). Therefore, $\ell \cong \ell_{\frac{a}{2^k}} \perp \ell_{\frac{d'}{2^k}}$. □

Note that if $b \equiv 0 \; mod \; (4)$ then $\ell \cong \ell_{\frac{a}{2^k}} \perp \ell_{\frac{d}{2^k}}$.

Suppose now that $\mathbb{Z}_{(2)} \otimes \ell_M$ is odd and $\varepsilon(M) = 0$. Then the diagonal entries of L are odd and the off-diagonal elements are odd multiples of $2^{-k}\alpha_1$. We may assume also that $r_2 \geqslant 4$, for otherwise $\mathbb{Z}_{(2)} \otimes \tau_M$ is cyclic. We may partition L as $L = \left(\begin{smallmatrix} E & F \\ F^{tr} & G \end{smallmatrix}\right)$, where $E \in GL(2, \mathbb{Z}/2^k\mathbb{Z})$, F is a $2 \times (r_1 - 4)$-submatrix with even entries and G is a $(r_2 - 4) \times (r_2 - 4)$-submatrix. Let $J = \left(\begin{smallmatrix} I_2 & -E^{-1}F \\ 0 & I_{r_2-4} \end{smallmatrix}\right)$. Then $\det(J) = 1$ and $J^{tr}LJ = \left(\begin{smallmatrix} E & 0 \\ 0 & G' \end{smallmatrix}\right)$, where $G' = G - F^{tr}E^{-1}F$. The columns of F are proportional, and the ratio $\frac{u_3}{u_4}$ is odd. Since the entries of F are odd multiples of $2^{-k}\alpha_1$ and since $E - I_2$ has even entries, $G' \equiv G \; mod \; (8)$. Iterating this process, we may replace L by a block-diagonal matrix, where the blocks are all 2×2 or 1×1, and are congruent *mod* (8) to the corresponding blocks of L. Each such 2×2 block is diagonalizable, by Lemma A8, and so we may easily represent ℓ_M as an orthogonal sum of pairings of rank 1.

If $\varepsilon(M) \neq 0$ then $\ell_M(q_i', s) = [2^{-k}\frac{\beta_i}{u_i}]$ and $\ell_M(s, s) = [2^{-k}z]$, where β_i, u_i and z are odd, and we first replace each q_i' by $\widetilde{Q}_i = q_i' - z^{-1}u^{-1}\beta_i s$. We then see that $\mathbb{Z}_{(2)} \otimes \ell_M \cong \ell_{2^{-k}z} \perp \widetilde{\ell}$, where the matrix for $\widetilde{\ell}$ has odd diagonal entries and even off-diagonal entries, and we may continue as before.

THEOREM A.9. *Let ℓ be a linking pairing on $(\mathbb{Z}/2^k\mathbb{Z})^\rho$. Then there is a Seifert manifold $M = M(0; S)$ such that the cone point orders are all powers of 2 and $\ell_M \cong \ell$. We may have either $\varepsilon(M) = 0$ or $\varepsilon(M) \neq 0$.*

PROOF. Suppose first that ℓ is even. Then ρ is also even, and $\ell \cong (E_0^k)^{\frac{\rho}{2}}$ or $(E_0^k)^{\frac{\rho}{2}-1} \perp E_1^k$. We shall choose r to be either $\rho + 2$ or $\rho + 1$, in order to construct examples with $\varepsilon(M) = 0$ or $\neq 0$.

Let $S = ((2^k, \beta_1), \ldots, (2^k, \beta_r))$ with $\beta_i = (-1)^i$ for $1 \leqslant i \leqslant r$. Then $\varepsilon(M) = -\frac{1}{2^k}$ if r is odd and $\varepsilon(M) = 0$ if r is even, and $\ell_M \cong (E_0^k)^r$.

If $\rho \equiv 2 \; mod \; (4)$ let $\beta_1 = -3$, $\beta_2 = \beta_3 = 1$ and $\beta_i = (-1)^i$ for $4 \leqslant i \leqslant r$. If $\rho \equiv 0 \; mod \; (4)$ let $\beta_1 = -5$, $\beta_2 = \cdots = \beta_5 = 1$ and $\beta_i = (-1)^i$ for $6 \leqslant i \leqslant r$. In each case, $\varepsilon(M) = -\frac{1}{2^k}$ if r is odd and $\varepsilon(M) = 0$ if r is even, and $\ell_M \cong (E_0^k)^{\frac{\rho}{2}-1} \perp E_1^k$.

Now suppose that ℓ is odd. Then $\ell \cong \perp_{i=1}^{\rho} \ell_{w_i}$ where $w_i = 2^{-k}b_i$ for $1 \leqslant i \leqslant \rho$. To construct an example with $\varepsilon(M) = 0$ we set $r = \rho + 2$ and $S = ((\alpha_1, \beta_1), \ldots, (\alpha_r, \beta_r))$, where $\alpha_1 = \alpha_2 = 2^{k+2}$, $\alpha_i = 2^k$ for $3 \leqslant i \leqslant \rho$, $\beta_2 = 1$, $\beta_i = 3b_i$ for $3 \leqslant i \leqslant r$ and $\beta_1 = -1 - 4\Sigma_{i=3}^r \beta_i$. Then $\varepsilon(M) = 0$ and $\ell_M \cong \ell$. (Here we may use the observation following Lemma A8.)

To construct an example with $\varepsilon_S \neq 0$ we set $r = \rho + 1$. Here we must take into account the change of basis suggested in the paragraph before the theorem. Suppose first that some $b_i \equiv \pm 3 \; mod \; (8)$. We may then arrange that $z = 3$ and the matrix for $\widetilde{\ell}$ is congruent *mod* (4) to a diagonal matrix. After reordering the summands, and allowing for a change of orientation, we may assume that $b_1 \equiv 3 \; mod \; (8)$. The we let $S = ((\alpha_1, \beta_1), \ldots, (\alpha_r, \beta_r))$, where $\alpha_1 = 2^{k+2}$, $\alpha_i = 2^k$ for $2 \leqslant i \leqslant r$, $\beta_2 = 1$, $\beta_i = 4 - b_i$ for $3 \leqslant i \leqslant r$ and $\beta_1 = 1 - 4\Sigma_{i=2}^r \beta_i$. Then $\varepsilon(M) = -\frac{1}{2^{k+2}}$ and $\ell_M \cong \ell$.

Finally, suppose that $b_1 \equiv \pm 1 \; mod \; (8)$ for all i. If $\rho = 1$ let $S = ((2^{k+1}, 1), (2^k, 1))$. If $\rho = 2$ and $b_1 \equiv -b_2 \equiv 1 \; mod \; (8)$, let $S = ((2^{k+1}, 1), (2^k, 1, (2^k, -1))$. Otherwise we may assume that $b_1 \equiv b_2 \ mod\,(8)$. But then $\ell_{w_1} \perp \ell_{w_2} \cong 2\ell_{w'}$, where $w' = 2^{-k}3$, and so we may use the construction in the preceding paragraph. □

The manifolds constructed in this section are either $\mathbb{H}^2 \times \mathbb{E}^1$-manifolds (if $\varepsilon(M) = 0$), or $\widetilde{\mathbb{SL}}$-manifolds (if $\varepsilon(M) \neq 0$), excepting only the flat manifold $M(0;(2,-1),(2,1),(2,-1),(2,1))$ and the $\mathbb{S}^3$-manifolds $M(0;(2,1),(2,1),(2,\beta))$ and $M(0;(4,1),(2,1),(2,\beta))$.

A.5 The General Case

Every linking pairing ℓ on a finite abelian group is realized by some oriented 3-manifold [**KK80**]. The next theorem suffices to show that there are pairings which cannot be realized by Seifert manifolds. We shall then show that there are no further constraints.

THEOREM A.10. *Let $M = M(g;S)$ be a Seifert manifold, and let $S(2) = ((\alpha_1,\beta_1),\dots,(\alpha_t,\beta_t))$ be the terms of the Seifert data with even cone point orders α_i. Then*

(1) *$\mathbb{Z}_{(2)} \otimes \ell_M$ has a non-trivial even component if and only if* $\omega = \max\{i \leqslant t \mid \frac{\alpha_1}{\alpha_i} \text{ is odd}\} \geqslant 3$;
(2) *the even component has exponent α_1, and all the other components are diagonalizable;*
(3) *if ω is even and $\varepsilon(M) \neq 0$ then the image of s in $\mathbb{Z}_{(2)} \otimes \tau_M$ generates a cyclic component of maximal exponent;*
(4) *if $\omega = 2$ then $\alpha_1\alpha_2\varepsilon_S$ is divisible by $4\alpha_3$ in $\mathbb{Z}_{(2)}$.*

PROOF. Block diagonalization does not change the parity of the entries in the diagonal blocks D_i. The first assertion then follows easily from the calculations of the first section. (More precisely, if ω is odd then $\alpha_1\varepsilon(M)$ is odd and $\mathbb{Z}_{(2)} \otimes \tau_M$ has exponent α_2. If moreover $\omega = 1$ then $\ell_M(s,s)$ is odd and (so) the homogeneous components are all odd. If $\omega > 1$ and is odd then $\ell_M(s,s)$ is even and the component of maximal exponent is even. If ω is even and $\varepsilon(M) = 0$ then the component of maximal exponent is non-trivial (and even) if and only if $\omega > 2$. If ω is even and $\varepsilon(M) \neq 0$ then s has order $> \alpha_1$, and $\ell_M(s,s)$ is odd. If moreover $\omega > 2$ then the component of exponent α_1 is even and has rank $\omega - 2$. In each case, all other components are odd.)

The final assertion is clear if $\varepsilon(M) = 0$ and follows by elementary arithmetic otherwise, for then $\alpha_1\varepsilon(M)$ is even. □

Theorem A10 provides a criterion for recognizing pairings which are not realizable by Seifert manifolds that is independent of the choice of generators. Let $N \cong N' \oplus N''$ be a finite abelian group, where the homogeneous summands of N' have exponent $\geqslant 4$ and $2N'' = 0$, and let $\ell = \ell' \perp \ell''$, where ℓ' is a pairing on N' and ℓ'' is a hyperbolic pairing on N''. Then $\ell(x,x) = 0$ for all $x \in N$ such that $2x = 0$. In particular, if N' is not cyclic then ℓ is not realizable by a Seifert manifold.

For example, if M is the $\mathbb{N}il^3$-manifold $M(0;(2,1),(2,1),(2,1),(2,-1))$ then $\ell_M \cong \ell_{\frac{1}{4}} \perp E_0^1$, but this pairing is not realizable by a Seifert manifold with $\varepsilon(M) = 0$. The pairing $E_0^2 \perp E_0^1$ is not realized by any Seifert manifold with orientable base orbifold. However, it is the linking pairing of $M(-2;((2,1),(2,1),(2,1),(2,1)))$.

If $\mathbb{Z}_{(2)} \otimes \tau_M$ has exponent divisible by 16 and a direct summand of order 2 then there are cone point orders α_1 and α_m such that α_1 is divisible by 4 and $\alpha_m = 2u_m$ with u_m odd, by Lemma 4.1. It then follows from Theorem 4.3 that $\mathbb{Z}_{(2)} \otimes \ell_M \cong \ell' \perp \ell_{\frac{1}{2}}$, for some pairing ℓ'. In particular, $E_0^4 \perp E_0^1$ is not realized by any orientable Seifert fibred 3-manifold at all.

The conditions of Theorem A10 are the only constraints on the class of linking pairings realized by Seifert manifolds with $\varepsilon(M) = 0$.

THEOREM A.11. *Let ℓ be a linking pairing on a finite abelian group A. Then there is a Seifert manifold $M = M(0;S)$ with $\varepsilon(M) = 0$ such that $\ell_M \cong \ell$ if and only if the components of the 2-primary summand of ℓ other than the component of maximal exponent are all odd.*

PROOF. The condition is necessary, by Theorem A10. By Lemma A1 and Theorem A4 it shall suffice to assume that A is 2-primary, and to realize ℓ by a Seifert manifold $M(0;S)$ with $\varepsilon(M) = 0$ and such that all the cone point orders are powers of 2.

Suppose that $\ell = \perp_{j=1}^t \ell_j$, where ℓ_j is a pairing on $(\mathbb{Z}/2^{k_j}\mathbb{Z})^{\rho_j}$, for $1 \leqslant j \leqslant t$, $k_1 > \cdots > k_t > 0$ and ℓ_j is odd for $2 \leqslant j \leqslant t$. We may assume that $t > 1$ and $\rho_j > 0$ for $1 \leqslant j \leqslant t$, since the homogeneous case is covered by Theorem A9.

We must have two cone points of order at least 2^{k_1}, and ρ_j further cone points of order 2^{k_j}, for $1 \leqslant j \leqslant t$. We shall set $\beta_2 = 1$ and choose each β_i with $i > 2$ compatibly with ℓ, essentially by induction on t. If we make these choices then we must have $\beta_1 = -\alpha_1 \Sigma_{i \geqslant 2} \frac{\beta_i}{\alpha_i}$. (Note that the presentation of ℓ_M given in the first section does not invoke β_1 when $\varepsilon(M) = 0$.)

Suppose first that ℓ_1 is even. Then ρ_1 is even and we must have $\alpha_1 = \alpha_2 = 2^{k_1}$ also. If ℓ_1 is hyperbolic let $\beta_i = (-1)^i$ for $3 \leqslant i \leqslant \rho_1 + 2$; if ℓ_1 is even but not hyperbolic and $\rho_1 \equiv 2$ *mod* (4) let $\beta_3 = 1$ and $\beta_i = (-1)^i$ for $4 \leqslant i \leqslant \rho_1 + 2$; and if ℓ_1 is even but not hyperbolic and $\rho_1 \equiv 0$ *mod* (4) let $\beta_3 = \beta_4 = \beta_5 = 1$ and $\beta_i = (-1)^i$ for $6 \leqslant i \leqslant \rho_1 + 2$.

When $p = 2$ block diagonalization does not change the parity of the entries of the blocks B_m and D_n. In our situation this process allows a more refined reduction. The block $DA = D_1$ has diagonal entries $-\beta_i(1+\beta_i)$ and off-diagonal entries $-\beta_i\beta_j$, for $3 \leqslant i,j \leqslant \rho_1 + 2$. Let $\Delta = -\mathrm{diag}[\beta_3, dots, \beta_{\rho_1+2}]$ and $N = D - \Delta$. Then $\Delta \equiv I$ *mod* (2), since the β_is are all odd, so $\Delta^2 \equiv I$ *mod* (4), and

$$N^2 \equiv -(\Sigma_{i=3}^{\rho_1+2}\beta_i^2)N \equiv \rho_1 N \quad mod\ (4).$$

In particular $N^2 \equiv 0$ *mod* (2), and so $D^{-1} \equiv D$ *mod* (2),. Since

$$[B_m^{tr} D B_n]_{pq} = -\beta_p\beta_q((\Sigma_{i=3}^{\rho_1+2}\beta_i^2)^2 + \Sigma_{i=3}^{\rho_1+2}\beta_i^3)$$

is even, for $3 \leqslant p,q \leqslant 2 + \Sigma_{j=1}^t \rho_j$, the first step of the reduction does not change the images of the complementary blocks *mod* (4). Truncating a geometric series gives

$D^{-1} \equiv D + (\rho_1 - 2)N \ mod\ (4)$. Hence the change $mod\ (8)$ depends only on the residues $mod\ (4)$ of the β_is, for $3 \leqslant i \leqslant \rho_1 + 2$, since $-4 \equiv 4\ mod\ (8)$. If $k_1 - k_2 \geqslant 2$ this step does not change the images $mod\ (8)$ at all.

The subsequent blocks have odd diagonal matrices and even off-diagonal elements. Nevertheless, a similar reduction applies, and further changes depend only on the β_is already determined. (If $k_j - k_{j+1} \geqslant 2$ for all j then they depend only on the ranks of the homogeneous terms.) It is clear that we may then choose the numerators β_i to realize all the pairings ℓ_j.

If ℓ_1 is odd then $\ell_1 = \perp_{i=1}^{\rho_1} \ell_{w_i}$, where $w_i = 2^{-k+1}b_i$ for $1 \leqslant i \leqslant \rho$. Let $\alpha_1 = \alpha_2 = 2^{k_1+2}$, and $\beta_i = 3b_i$ for $3 \leqslant i \leqslant \rho_1 + 2$. Then we may continue as before. □

A similar argument applies when $\varepsilon(M) \neq 0$. However we shall only sketch the argument in this case.

THEOREM A.12. *Let ℓ be a linking pairing on a finite abelian group A. Then there is a Seifert manifold $M = M(0;S)$ with $\varepsilon(M) \neq 0$ such that $\ell_M \cong \ell$ if and only if the 2-primary components of ℓ satisfy the conditions of Theorem A10.*

PROOF. Suppose first that A is 2-primary, and that $\ell = \perp_{j=1}^{t} \ell_j$, where ℓ_j is a pairing on $(\mathbb{Z}/2^{k_j}\mathbb{Z})^{\rho_j}$, for $1 \leqslant j \leqslant t$. Suppose also that $k_1 > \cdots > k_t > 0$ and ℓ_j is odd, for $3 \leqslant j \leqslant t$, and that either ℓ_2 is odd or $\rho_1 = 1$ and ℓ_2 is even. We may again assume that $t > 1$ and $\rho_j > 0$, for $1 \leqslant j \leqslant t$, since the homogeneous case is covered by Theorem A9.

The cone point orders are essentially determined by Theorem A10. If ℓ_1 is even we must have $\rho_1 + 1$ cone points of order 2^{k_1} and ρ_j cone points of order 2^{k_j}, for $2 \leqslant j \leqslant t$. If ℓ_2 is even then $\rho_1 = 1$ and we must have $\rho_2 + 2$ cone points of order 2^{k_2} and ρ_j cone points of order 2^{k_j}, for $3 \leqslant j \leqslant t$. If the components ℓ_j are all odd then we may choose one cone point of order $\alpha_1 > 2^{k_1}$ and ρ_j cone points of order 2^{k_j}, for $1 \leqslant j \leqslant t$. (If $\rho_1 = 1$ and $k_1 - k_2 \geqslant 2$ we could instead choose two cone points of order 2^{k_0}, where $k_1 > k_0 > k_2$, and ρ_j cone points of order 2^{k_j}, for $2 \leqslant j \leqslant t$. However, we shall not use this option.)

We then have $\varepsilon(M) = \frac{b}{d}$, where b is odd and $d = \alpha_1$ if ℓ_2 is odd and $d = 2^{2k_2 - k_1}$ if ℓ_2 is even. Let $\beta_2 = 1$ and choose the β_is with $i > 2$ compatibly with ℓ and our choice for $\varepsilon(M)$. Let $\Sigma' = \Sigma_{i \geqslant 2} \frac{\beta_i}{\alpha_i}$. Then we must have $\beta_1 = -\alpha_1(\varepsilon(M) + \Sigma')$. (In each case this is odd.)

From here the strategy is as in Theorem A11, and we shall not give further details for this part of the construction.

In order to construct M we may assume that $b_1 = 1$, and hence that $\varepsilon(M) = \frac{1}{2^k}$. However, when A also has odd primary summands we cannot use Lemma A1 to reduce to the 2-primary case. In order to extend the argument of Theorem A5 to pairings with 2-primary summands, it is convenient to allow b to be the *inverse* of an odd integer.

Let P be the finite set of odd primes dividing the order of A. Let $\Pi = \Pi_{p \in P} p$ and E be the product of the exponents of the odd primary summands of A, and let $\varepsilon(M) = \frac{1}{dE\Pi}$.

If ℓ_2 is odd we may define Seifert data $S(p)$ as in Theorem A5, for each $p \in P$. We then construct Seifert data $S(2) = ((\alpha_2, \beta_2), \dots, (\alpha_{\rho(2)}, \beta_{\rho(2)}))$ with α_i a power of 2, for $i > 1$, as above. However, we now replace α_1 by $\widetilde{\alpha} = \alpha_1 E\Pi$.

If ℓ_2 is even then for each $p \in P$ we have $\widetilde{\alpha}\varepsilon(M) \equiv \frac{\alpha_1}{d}$ *mod* (p), and we must modify the choices of some of the $\beta_i^{(p)}$s for the Seifert data $S(p)$ slightly.

Finally, let $\widetilde{\beta} = -\widetilde{\alpha}(\varepsilon(M) + \Sigma' + \Sigma_{p\in P}\Sigma_{i=1}^{t(p)} \frac{\beta_i^{(p)}}{\alpha_i^{(p)}})$, where d and Σ' are defined as before, in terms of the Seifert data of the cone points of even order. (Note that $\widetilde{\beta}$ is odd and $\widetilde{\beta} \equiv -1$ *mod* (p), for $p \in P$, and $\varepsilon(M) = \frac{2^m}{\widetilde{\alpha}}$, for some $m \geqslant 0$.) □

Appendix B Homologically Balanced Nilpotent Groups

In this appendix we shall present what we know about homologically balanced nilpotent groups. If G is such a group it can be generated by three elements [**Lub83**, Theorem 2.7], and if $\beta_1(G;\mathbb{Q}) = 3$ then $G \cong \mathbb{Z}^3$ [**Hi22**].

B.1 Unipotent Actions

We shall extend the term "unipotent", to say that an action $\alpha : G \to Aut(A)$ is unipotent if $\alpha(g)$ is unipotent for all $g \in G$.

LEMMA B.1. *Let N be a finitely generated nilpotent group which acts unipotently on a finitely generated abelian group A, and let $\mathfrak{n}$ be the augmentation ideal of $\mathbb{Z}[N]$. Then A has a finite filtration $A = A_1 > \cdots > A_k = A^N > A_{k+1} = 0$ by $\mathbb{Z}[N]$-submodules, where A^N is the fixed subgroup and $\mathfrak{n}A_i \leqslant A_{i+1}$, for $i \leqslant k$.*

PROOF. We induct on the length of the upper central series of N. The centre ζN is a non-trivial abelian group which acts unipotently on A, and it is easy to see that $A^{\zeta N} \neq 0$. The quotient $N/\zeta N$ acts unipotently on each of $A^{\zeta N}$ and $\overline{A} = A/A^{\zeta N}$, and so these each have such filtrations, by the inductive hypothesis. The pre-images of the filtration of $\overline{A}$ in A combine with the filtration of $A^{\zeta N}$ to give the required filtration. □

It is easy to see that the product of commuting unipotent automorphisms is unipotent. This observation extends to show that an action of a nilpotent group N is unipotent if N is generated by elements which act unipotently.

We shall find the following notion useful in many of our arguments. Let G be a group and F a field. Then an $F[G]$-module V is *canonically subsplit* if it contains a non-trivial direct sum of $F[G]$-submodules. If G acts unipotently on V and V is canonically subsplit then the subspaces of the summands fixed by G/K are non-trivial, by Lemma B1, and so the subspace V^G fixed by G has dimension > 1.

LEMMA B.2. *Let A be a finitely generated abelian group and p a prime such that A has non-trivial p-torsion and $\dim_{\mathbb{F}_p} A/pA > 1$. If p is odd or if $p = 2$ and A has no $\mathbb{Z}/2\mathbb{Z}$ summand then $H_2(A;\mathbb{F}_p)$ and $H^2(A;\mathbb{F}_p)$ are each canonically subsplit with respect to the natural action of (subgroups of)* $\mathrm{Aut}(A)$.

PROOF. Let $W = (A/pA) \wedge (A/pA)$ and $A^* = Hom(A;\mathbb{F}_p) = H^1(A;\mathbb{F}_p)$.

Then there is a natural splitting $H_2(A;\mathbb{F}_p) = W \oplus Tor(A, \mathbb{F}_p)$ if p is odd [**Bro**, Chapter V.6], or if $p = 2$ and A has no $\mathbb{Z}/2\mathbb{Z}$ summand [**IZ18**]. There is also a natural epimorphism $\theta : H^2(A;\mathbb{F}_p) \to Hom(W, \mathbb{F}_p)$, with kernel isomorphic to $Ext(A;\mathbb{F}_p)$ [**Bro**, Exercises IV.3.8 and V.6.5].

If p is odd then cup product induces a monomorphism $c_A : A^* \wedge A^* \to H^2(A;\mathbb{F}_p)$, since A is abelian. If $p = 2$ then cup product defines a homomorphism from $A^* \odot A^*$ to $H^2(A;\mathbb{F}_2)$. Since A has no $\mathbb{Z}/2\mathbb{Z}$ summand, $Sq(a) = a \cup a = 0$ for all $a \in A^*$, and so cup product again induces a monomorphism $c_A : A^* \wedge A^* \to H^2(A;\mathbb{F}_p)$ [**Hi87**]. It is easily seen from the formulae in [**Bro**] that $\theta \circ c_A$ is an isomorphism, and so $H^2(A;\mathbb{F}_p)$ is naturally isomorphic to $(A^* \wedge A^*) \oplus Ext(A;\mathbb{F}_p)$.

The summands are all non-trivial, since A has non-trivial p-torsion and A/pA is not cyclic. Thus $H_2(A;\mathbb{F}_p)$ and $H^2(A;\mathbb{F}_p)$ are each canonically subsplit. □

The case when A has a summand of exponent 2 seems more complicated, and we consider only the cohomology.

LEMMA B.3. *Let A be a finitely generated abelian group with a non-trivial summand of exponent 2 and such that $\dim_{\mathbb{F}_2} A/2A > 1$. Suppose that a finitely generated nilpotent group N acts unipotently on A. Then $\dim_{\mathbb{F}_2} H^2(A;\mathbb{F}_2)^N > 1$.*

PROOF. We may assume that $A \cong B \oplus E$, where $E \cong (\mathbb{Z}/2\mathbb{Z})^s \neq 0$ and B has no summand of order 2. The subspace B^* of $A^* = Hom(A, \mathbb{F}_2) = H^1(A;\mathbb{F}_2)$ consisting of homorphisms which factor through homomorphisms to $\mathbb{Z}/4\mathbb{Z}$ is canonical. Clearly $B^* \cong Hom(B, \mathbb{F}_2)$ and $A^*/B^* \cong E^* = Hom(E, \mathbb{F}_2)$. Hence $A^* \cong B^* \oplus E^*$, but this splitting is not canonical. Cup product induces a homomorphism $c_A : A^* \odot A^* \to H^2(A;\mathbb{F}_2)$, with kernel $2A/4A \cong B^*$, since A is abelian [**Hi87**]. There is also a natural squaring map $Sq : A^* \to H^2(A;\mathbb{F}_2)$ with kernel B^*.

If $B = 0$ then A is an elementary 2-group and $A^* = E^*$, and c_A is a monomorphism. Let $A_1 > \cdots > A_{k+1} = 0$ be a filtration of A^* by $\mathbb{F}_p[N]$-submodules, as in Lemma B.1. Then $A_k \odot A_k$ is fixed by N. If $\dim_{\mathbb{F}_2} A_k > 1$ then $\dim_{\mathbb{F}_2} A_k \odot A_k \geqslant 3$. If A_k has dimension 1, and is generated by b then $b \odot b$ is fixed by N. If $a \in A_{k-1}$ then each element of N either fixes a or sends it to $a+b$. In either case $a \odot (a+b)$ is fixed by N. Since $\dim_{\mathbb{F}_2} A_{k-1} \geqslant 2$ the subspace generated by $\{a \odot (a+b) \mid a \in A_{k-1}\} \cup \{b \odot b\}$ is fixed by N, and so $\dim_{\mathbb{F}_2} H^2(A;\mathbb{F}_2)^N > 1$.

The images of $B^* \otimes A^*$ and $Sq(A^*) = Sq(E^*)$ are canonical submodules of $H^2(A;\mathbb{F}_2)$, with trivial intersection. Hence they are invariant under the action of automorphisms of A, and so if $B \neq 0$ then we again have $\dim_{\mathbb{F}_2} H^2(A;\mathbb{F}_2)^N > 1$. □

COROLLARY. *Let A be a finitely generated abelian group, ψ be a unipotent automorphism of A, and p be a prime. If A has non-trivial p-torsion and $\dim_{\mathbb{F}_p} A/pA > 1$ then $\dim_{\mathbb{F}_p} \mathrm{Ker}(H_2(\psi;\mathbb{F}_p) - I) = \dim_{\mathbb{F}_p} \mathrm{Ker}(H^2(\psi;\mathbb{F}_p) - I) > 1$.*

PROOF. Let N be the cyclic subgroup of $Aut(A)$ generated by ψ. We shall write $H_i(\psi)$ and $H^j(\psi)$ instead of $H_i(\psi;\mathbb{F}_p)$ and $H^j(\psi;\mathbb{F}_p)$, for simplicity of notation. Then $H_i(A;\mathbb{F}_p)^N = \mathrm{Ker}(H_i(\psi) - I)$ and $H^j(A;\mathbb{F}_p)^N = \mathrm{Ker}(H^j(\psi) - I)$, for any i. If φ is an endomorphism of a finite dimensional vector space V then $\dim \mathrm{Cok}(\varphi) = \dim \mathrm{Ker}(\phi)$ and if φ^* is the induced endomorphism of the dual vector space V^* then

φ^* and φ have the same rank. Hence the corollary follows from Lemma B2, if p is odd, and from Lemma B3, if $p = 2$. □

It does not seem obvious that $\dim_{\mathbb{F}_p} H_2(A;\mathbb{F}_p)^N$ and $\dim_{\mathbb{F}_p} H^2(A;\mathbb{F}_p)^N$ are equal when N is not cyclic.

If $\dim_{\mathbb{F}_p} A/pA \geqslant 4$ then the restriction of $H_2(\psi;\mathbb{F}_p) - I$ to $(A/pA) \wedge (A/pA)$ has kernel of dimension > 1, and so $\dim_{\mathbb{F}_p} \mathrm{Ker}(H_2(\psi;\mathbb{F}_p) - I) > 1$. In [**Hi22**] a related observation for free abelian groups of rank $\geqslant 4$ is used to show that if G is a metabelian nilpotent group with $h(G) > 4$ then $\beta_2(G;\mathbb{Q}) > \beta_1(G;\mathbb{Q})$. One of the difficulties in extending the approach of this paper to more general nilpotent groups is the lack of an analogue to the above lemmas for non-abelian p-groups.

B.2 Virtually Cyclic Groups: $h \leqslant 1$

A finitely generated nilpotent group G is finite if and only if $h(G) = 0$, equivalently, if and only if $\beta_1(G;\mathbb{Q}) = 0$. The Sylow subgroups of a finite nilpotent group G are characteristic, and G is the direct product of its Sylow subgroups [**Rob**, 5.2.4]. It then follows from the Künneth Theorem that $H_2(G) = 0$ if and only if $H_2(P) = 0$ for all such Sylow subgroups P. On the other hand, it is not clear that if $H_2(G) = 0$ then G must have a balanced presentation, even if this is so for each of its Sylow subgroups. (The examples in [**HW85**] of finite perfect groups with trivial multiplicator but without balanced presentations are not nilpotent.)

If p is an odd prime then every 2-generator metacyclic p-group P with $H_2(P) = 0$ has a balanced presentation

$$\langle a, b \mid b^{p^{r+s+t}} = a^{p^{r+s}},\ bab^{-1} = a^{1+p^r} \rangle,$$

where $r \geqslant 1$ and $s, t \geqslant 0$. (The order of such a group is $p^{3r+2s+t}$.) There are other metacyclic 2-groups and other p-groups with 2-generator balanced presentations. A handful of 3-generated p-groups (for $p = 2$ and 3) are also known to have balanced presentations. (See [**HNO'B96**] for a survey of what was known in the mid-1990s.)

If T is one of the 2-generator metacyclic p-groups of the first section, then $H^2(T;\mathbb{F}_p)$ has no canonically split subspace, and such groups do arise as the torsion subgroups of homologically balanced nilpotent groups G with $h(G) = 1$. Can we at least use such an argument to show that the torsion subgroup must be homologically balanced?

We include the following simple lemma as some of the observations are not explicit in our primary reference [**Rob**].

LEMMA B.4. *Let G be a finitely generated nilpotent group, and let T be its torsion subgroup. Then the following are equivalent*

(1) $\beta_1(G;\mathbb{Q}) = 1$;
(2) $h(G) = 1$;
(3) $G/T \cong \mathbb{Z}$;
(4) $G \cong T \rtimes_\psi \mathbb{Z}$, *where ψ is an automorphism of T.*

PROOF. In each case G is clearly infinite, and so there is an epimorphism $f : G \to \mathbb{Z}$. Since G is finitely generated, so is $K = \mathrm{Ker}(f)$. If $\beta_1(G;\mathbb{Q}) = 1$ then K

is finite, by Lemma 9.2. If $h(G) = 1$ then $h(K) = 0$, so K is again finite. In each case, $K = T$ and $G/T \cong \mathbb{Z}$. If $G/T \cong \mathbb{Z}$ and $t \in G$ represents a generator of G/T then conjugation by t defines an automorphism ψ of T, and $G \cong T \rtimes_\psi \mathbb{Z}$. Finally, it is clear that (4) implies each of (1) and (2). □

We could also describe the groups considered on this lemma as the nilpotent groups which are virtually $\mathbb{Z}$, and as the nilpotent groups with two ends.

In the next lemma we do not assume that G is nilpotent.

LEMMA B.5. *Let $G \cong T \rtimes_\psi \mathbb{Z}$ be a homologically balanced group, where T is finite. Then $H_2(G)$ is a finite cyclic group, and its order is divisible by the order of the torsion subgroup of G^{ab}.*

PROOF. It is immediate from the Wang sequence for the integral homology of G as an extension of $\mathbb{Z}$ by T that $H_2(G)$ is finite. It is also clear that $C = \mathrm{Cok}(\psi^{ab} - I)$ is the torsion subgroup of G^{ab}. Since T is finite, $|\mathrm{Ker}(\psi^{ab} - I)| = |\mathrm{Cok}(\psi^{ab} - I)|$, and so $|C|$ divides the order of $H_2(G)$.

If p is a prime then $\dim_{\mathbb{F}_p} Tor(\mathbb{F}_p, G^{ab}) = \dim_{\mathbb{F}_p} Tor(\mathbb{F}_p, C) = \beta_1(G; \mathbb{F}_p) - 1$, since $G^{ab} \cong \mathbb{Z} \oplus C$. Therefore

$$\dim_{\mathbb{F}_p} Hom(H_2(G), \mathbb{F}_p) = \beta_2(G; \mathbb{F}_p) - \beta_1(G; \mathbb{F}_p) + 1,$$

by the Universal Coefficient exact sequences of Chapter 1. Since G is homologically balanced, this is at most 1, for all primes p, and so $H_2(G)$ is cyclic. □

If G is nilpotent then $H_2(\psi) - I$ is a nilpotent endomorphism of $H_2(T)$, and so C and $H_2(G)$ have the same order if and only if $H_2(T) = 0$.

THEOREM B.6. *Let $G \cong T \rtimes_\psi \mathbb{Z}$, where T is a finite nilpotent group and ψ is a unipotent automorphism of T. If G is homologically balanced then*

(1) *G can be generated by 2 elements;*
(2) *if the Sylow p-subgroup of T is abelian then it is cyclic;*
(3) *if T is abelian then $G \cong \mathbb{Z}/m\mathbb{Z} \rtimes_n \mathbb{Z}$, for some $m, n \neq 0$ such that m divides a power of $n - 1$.*

PROOF. Let p be a prime. Then $\dim_{\mathbb{F}_p} H_1(T; \mathbb{F}_p) = \dim_{\mathbb{F}_p} Tor(T^{ab}, \mathbb{F}_p)$, since T is finite. Moreover, $\psi^{ab} - I$ and $Tor(\psi^{ab}, \mathbb{F}_p) - I$ have the same rank. Since $Tor(T^{ab}, \mathbb{F}_p)$ is a natural quotient of $H_2(T; \mathbb{F}_p)$, exactness of the Wang sequence implies that $\beta_2(G; \mathbb{F}_p) \geqslant 2(\beta_1(G; \mathbb{F}_p) - 1)$. Since G is homologically balanced, $\beta_1(G; \mathbb{F}_p) \leqslant 2$. Hence the p-torsion of G^{ab} is cyclic. Therefore, $G^{ab} \cong \mathbb{Z} \oplus C$ for some finite cyclic group. Since G is nilpotent and G^{ab} can be generated by two elements, so can G.

The Sylow subgroups of T are characteristic, and ψ restricts to a unipotent automorphism of each such subgroup. Suppose that the Sylow p-subgroup of T is an abelian group A. Since $H^2(A; \mathbb{F}_p) = Hom(H_2(A; \mathbb{F}_p), \mathbb{F}_p)$, the endomorphisms $H^2(\psi; \mathbb{F}_p) - I$ and $H_2(\psi; \mathbb{F}_p) - I$ have the same rank. Hence

$$\dim(\mathrm{Ker}(H^2(\psi; \mathbb{F}_p) - I) = \dim(\mathrm{Ker}(H_2(\psi; \mathbb{F}_p) - I) = \dim(\mathrm{Cok}(H_2(\psi; \mathbb{F}_p) - I) \leqslant 1.$$

Hence A must be cyclic, by the corollary to Lemma B3.

It follows immediately that if T is abelian then it is a direct product of cyclic groups of relatively prime orders, and so is cyclic. Hence $G \cong \mathbb{Z}/m\mathbb{Z} \rtimes_n \mathbb{Z}$, for some $m \geqslant 1$ and n such that $(m, n) = 1$. Such a semi-direct product is nilpotent if and only if m divides some power of $n - 1$. □

Every semi-direct product $\mathbb{Z}/m\mathbb{Z} \rtimes_n \mathbb{Z}$ has a balanced presentation

$$\langle a, t \mid a^m = 1,\ tat^{-1} = a^n \rangle.$$

The simplest examples with T non-abelian are the groups $Q(8k) \rtimes \mathbb{Z}$, with the balanced presentations $\langle t, x, y \mid x^{2k} = y^2,\ tx = xt,\ tyt^{-1} = xy \rangle$, which simplify to

$$\langle t, y \mid [t, y]^{2k} = y^2,\ [t, [t, y]] = 1 \rangle.$$

Let $m = p^s$, where p is a prime and $s \geqslant 1$, and let G be the group with presentation

$$\langle t, x, y \mid txt^{-1} = y,\ tyt^{-1} = x^{-1}y^2,\ yxy^{-1} = x^{m+1} \rangle.$$

Conjugating the final relation with t gives $x^{-1}yx = y^{m+1}$, and we see that the torsion subgroup T has presentation $\langle x, y \mid x^m = y^m,\ yxy^{-1} = x^{m+1} \rangle$, and so is one of the metacyclic p-groups mentioned above. Moreover, G is nilpotent, $\zeta G = \langle x^m \rangle$ and $G' = \langle x^m, x^{-1}y \rangle$ is abelian. Hence G is metabelian. Each of the groups that we have described here is 2-generated and its torsion subgroup is homologically balanced.

B.3 Metabelian Nilpotent Groups with Hirsch Length > 1

All known examples of nilpotent groups with balanced presentations and Hirsch length $h > 1$ are torsion-free. We have not yet been able to show that this must be so. However, if such a group is also metabelian, but not $\mathbb{Z}^3$, then $h(G) \leqslant 4$ and $\beta_1(G; \mathbb{Q}) = 2$ [**Hi22**, Theorems 7 and 15]. Our main result implies that there are just three such groups with $G/G' \cong \mathbb{Z}^2$. The argument again involves finding normal subgroups with "large enough" homology to affect the Betti numbers of the group. We develop a number of lemmas to this end. (The Lyndon-Hochschild-Serre spectral sequences for the homology and cohomology of a group which is an extension of $\mathbb{Z}^2$ by a normal subgroup shall largely replace the Wang sequences used earlier.)

LEMMA B.7. *Let G be a finitely generated nilpotent group, and let T be its torsion subgroup. Then the following are equivalent:*

(1) *$\beta_1(G; \mathbb{Q}) = 2$ and $\beta_2(G; \mathbb{Q}) = 1$;*
(2) *$h(G) = 2$;*
(3) *$G/T \cong \mathbb{Z}^2$.*

If these conditions hold and G is homologically balanced then $H_2(G) \cong \mathbb{Z} \oplus \mathbb{Z}/e\mathbb{Z}$, for some $e \geqslant 1$.

PROOF. If (1) holds then $h(G) \geqslant 2$, and so $h(G) = 2$, by the corollary to Lemma 9.2. It is easy to see that (2) and (3) are equivalent, and imply (1). The final assertion follows from Corollary 9.2.1 and Lemma 9.3. □

We could also describe the groups considered in this lemma as the nilpotent groups which are virtually $\mathbb{Z}^2$.

LEMMA B.8. *Let F be a field and A be a finite dimensional $F[\mathbb{Z}^2]$-module, and let $b_i = \dim_F H_i(\mathbb{Z}^2; A)$, for $i \geqslant 0$. Then $b_2 = b_0$ and $b_1 = b_0 + b_2 = 2b_0$.*

PROOF. We may compute $H_i(\mathbb{Z}^2; A) = Tor_i^{F[\mathbb{Z}^2]}(F, A)$ from the complex

$$0 \to A \to A^2 \to A \to 0,$$

in which the differentials are $\partial^1 = \begin{bmatrix} (x-1)id_A \\ (y-1)id_A \end{bmatrix}$, and $\partial^2 = \begin{bmatrix} (y-1)id_A, (1-x)id_A \end{bmatrix}$ where $\{x, y\}$ is a basis for $\mathbb{Z}^2$. Since the matrix for ∂^2 is the transpose of that for ∂^1 (up to a change of sign in the second block), they have the same rank. Hence $b_2 = \dim_F \mathrm{Ker}(\partial_2) = \dim_F \mathrm{Cok}(\partial_1) = b_0$. The final assertion follows since $b_0 - b_1 + b_2 = 1 - 2 + 1 = 0$ is the Euler characteristic of the complex. □

The modules $H_2(\mathbb{Z}^2; A)$ and $H_0(\mathbb{Z}^2; A)$ are the submodule of fixed points and the coinvariant quotient modules of the $\mathbb{Z}^2$-action, respectively. Minor adjustments give similar results for $\dim_F H^j(\mathbb{Z}^2; A)$. (We may also use Poincaré duality for $\mathbb{Z}^2$ to relate homology and cohomology.)

Recall that if K is a normal subgroup of a group G then conjugation in G induces a natural action of G/K on the homology and cohomology of K.

LEMMA B.9. *Let G be a finitely generated nilpotent group with a normal subgroup K such that $G/K \cong \mathbb{Z}^2$, and suppose that $\beta_2(G; F) \leqslant \beta_1(G; F)$ for some field F. Then $\dim_F H^2(K; F)^{G/K} \leqslant 1$.*

PROOF. We note first that $\beta_1(G; F) = 2$ or 3, since $\beta_2(G; F) \leqslant \beta_1(G; F)$ [**Lub83**, Theorem 2.7]. We may assume that $A = H^1(K; F) \neq 0$, since G is nilpotent. Let $N = G/K$ and $b_i = \dim_F H^i(N; A)$. The LHS spectral sequence for cohomology with coefficients F for G as an extension of N by K gives two exact sequences

$$0 \to H^1(N; F) \to H^1(G; F) \to A^N \xrightarrow{d_2^{0,1}} H^2(N; F) \to H^2(G; F) \to J \to 0$$

and

$$0 \to H^1(N; A) \to J \to H^2(K; F)^N \xrightarrow{d_2^{0,2}} H^2(N; A).$$

The first sequence gives $\dim_F J \leqslant \beta_2(G; F)$. Then $b_1 = b_0 + b_2 = 2b_0$, by Lemma B8, and $b_0 > 0$, since $A \neq 0$. Hence $b_0 - b_1 < 0$, and so the second sequence gives $\dim_F H^2(K; F)^N \leqslant \beta_2(G; F) + b_0 - b_1 \leqslant \beta_2(G; F) - 1$. In particular, $\dim_F H^2(K; F)^N \leqslant 1$ if $\beta_2(G; F) = 2$.

If $\beta_2(G; F) = 3$ then $\beta_1(G; F) = 3$ also, by Corollary 9.2.1. Therefore $\dim_F \mathrm{Ker}(d_2^{0,1}) = 1$. If $d_2^{0,1} \neq 0$ then $b_0 = 2$ and so $b_1 = 4$, by Lemma B8. But then $\beta_2(G; F) \geqslant 4$. Therefore $d_2^{0,1} = 0$, and so $b_0 = 1$. Hence $b_1 = 2$ and $b_2 = 1$, and $d_2^{0,2}$ is a monomorphism. Hence we again have $\dim_F H^2(K; F)^N \leqslant 1$. □

In particular, $H^2(K; F)$ is not canonically subsplit. A parallel argument using homology shows that $\dim_F H_0(G/K; H_2(K; F)) \leqslant 1$.

LEMMA B.10. *Let P be a non-trivial finite p-group and $K \cong \mathbb{Z} \times P$. Then $H^2(K;\mathbb{F}_p)$ is canonically subsplit.*

PROOF. We shall use the Universal Coefficient exact sequence for cohomology. The projection of K onto $K/P \cong \mathbb{Z}$ determines (up to sign) a class $\eta \in H^1(K;\mathbb{F}_p) = Hom(K,\mathbb{F}_p)$, and cup product with η maps $H^1(K;\mathbb{F}_p)$ non-trivially to $H^2(K;\mathbb{F}_p)$, by the Künneth Theorem. The restriction from $Ext(K^{ab},\mathbb{F}_p)$ to $Ext(P^{ab},\mathbb{F}_p)$ is an isomorphism, and so $Ext(K^{ab},\mathbb{F}_p)$ and $\eta \cup H^1(K;\mathbb{F}_p)$ have trivial intersection. Hence $Ext(K^{ab},\mathbb{F}_p) \oplus (\eta \cup H^1(K;\mathbb{F}_p))$ is a subspace of $H^2(K;\mathbb{F}_p)$, and the summands are invariant under the action of automorphisms of K, by the naturality of the Universal Coefficient Theorem. The summands are non-trivial, since $P \neq 1$. □

The next four lemmas (leading up to Theorem B15) consider nilpotent groups which are extensions of $\mathbb{Z}^2$ by finite normal subgroups.

LEMMA B.11. *Let G be a finitely generated nilpotent group, and let T be its torsion subgroup. Let P be a non-trivial Sylow p-subgroup of T and let $\gamma_p : G \to Out(P)$ be the homomorphism determined by conjugation in G. If $G/T \cong \mathbb{Z}^2$ and the image of γ_p is cyclic then $\beta_2(G;\mathbb{F}_p) > \beta_1(G;\mathbb{F}_p)$.*

PROOF. We may write $G \cong K \rtimes_\psi \mathbb{Z}$, where ψ is a unipotent automorphism of K, and K is in turn an extension of $\mathbb{Z}$ by T. Let P be the Sylow p-subgroup of T, and let N be the product of the other Sylow subgroups of T. Since the Sylow subgroups of T are characteristic, conjugation in G determines a homomorphism $\gamma_p : G \to Out(P)$. Moreover, N is normal in G, and the projection of G onto G/N induces isomorphisms on homology and cohomology with coefficients $\mathbb{F}_p$. Hence we may assume that $N = 1$ and so $T = P$ is a non-trivial p-group.

If the image of γ_P is cyclic then γ_P factors through an epimorphism $f : G \to \mathbb{Z}$, with kernel $K \cong \mathbb{Z} \times P$. Since $H^2(K;\mathbb{F}_p)$ has a subspace which is the direct sum of non-trivial canonical summands, by Lemma B10, $\dim_{\mathbb{F}_p} \mathrm{Ker}(H^2(\psi;\mathbb{F}_p) - I) > 1$ (as in Lemma B2). The result now follows from Lemma 9.2. □

Thus the group with presentation $\langle x, y \mid [x,[x,y]] = [y,[x,y]] = [x,y]^p = 1\rangle$ mentioned above does not have a balanced presentation. Similarly, no nilpotent extension of $\mathbb{Z}^2$ by $Q(8)$ can have a balanced presentation, since the abelian subgroups of $Out(Q(8)) \cong S_3$ are cyclic.

If p is an odd prime and C is a cyclic p-group then $\mathrm{Aut}(C)$ is cyclic, and so Lemma B1 may apply. However, dealing with 2-torsion again requires more effort.

LEMMA B.12. *Let G be a finitely generated nilpotent group, and let T be its torsion subgroup. If $G/T \cong \mathbb{Z}^2$ and the Sylow 2-subgroup of T is a non-trivial cyclic group then $\beta_2(G;\mathbb{F}_2) > \beta_1(G;\mathbb{F}_2)$.*

PROOF. We may factor out the maximal odd-order subgroup of T without changing the $\mathbb{F}_2$-homology, and so we may assume that $T \cong \mathbb{Z}/k\mathbb{Z}$, where $k = 2^n$, for some $n \geqslant 1$. We may also assume that the action of G on T by conjugation does not factor through a cyclic group, by Lemma B10, and so $k \geqslant 8$. Let U be the subgroup of $(\mathbb{Z}/k\mathbb{Z})^\times$ represented by integers $\equiv 1\ mod\ (4)$. Then $Aut(\mathbb{Z}/k\mathbb{Z}) \cong \{\pm 1\} \times U$.

It is easily verified that non-cyclic subgroups of $Aut(\mathbb{Z}/k\mathbb{Z})$ have $\{\pm 1\}$ as a direct factor, and so G has a presentation

$$\langle x, y, z \mid [x, y] = z^f,\ z^k = 1,\ xzx^{-1} = z^{-1},\ yzy^{-1} = z^\ell\rangle,$$

where f divides k, $1 < \ell < k$ and $\ell \equiv 1 \ mod\ (4)$. Let m be a mutiplicative inverse for $\ell\ mod\ (k)$, so that $1 < m < k$ and $m\ell = wk + 1$ for some $w \in \mathbb{Z}$. Note that $\beta_1(G; \mathbb{F}_2) = 2$ if $f = 1$ and is 3 if $f > 1$.

The ring $\mathbb{Z}[G]$ is a twisted polynomial extension of $\mathbb{Z}[\mathbb{Z}/k\mathbb{Z}] = \mathbb{Z}[z]/(z^k - 1)$, and so is noetherian. We may assume that each monomial is normalized in alphabetical order: $x^h y^i z^j$, for exponents $h, i \in \mathbb{Z}$ and $0 \leqslant j < k$. Let $\nu = \Sigma_{i=0}^{k-1} z^i$ be the norm element for $\mathbb{Z}[\mathbb{Z}/k\mathbb{Z}]$. Then $z\nu = \nu$, so $\nu^2 = k\nu$, and ν is central in $\mathbb{Z}[G]$. We shall use the fact that if $\gamma, \delta \in \mathbb{Z}[G]$ are such that $\gamma\nu = 0$ and $\delta(z - 1) = 0$ then $\gamma = \gamma'(z - 1)$ and $\delta = \delta'\nu$, for some $\gamma', \delta' \in \mathbb{Z}[G]$. On the other hand, non-zero terms not involving z are not zero-divisors in $\mathbb{Z}[G]$.

The augmentation module $\mathbb{Z}$ has a Fox-Lyndon partial resolution

$$C_2 \xrightarrow{\partial_2} C_1 \xrightarrow{\partial_1} C_0 = \mathbb{Z}[G] \xrightarrow{\varepsilon} \mathbb{Z} \to 0,$$

where $\varepsilon : \mathbb{Z}[G] \to \mathbb{Z}$ is the augmentation homomorphism, $C_1 \cong \mathbb{Z}[G]^3$ has basis $\{e_x, e_y, e_z\}$ corresponding to the generators and $C_2 \cong \mathbb{Z}[G]^4$ has basis $\{r, s, t, u\}$ corresponding to the relators $r = z^f yxy^{-1}x^{-1}$, $s = z^k$, $t = zxzx^{-1}$ and $u = z^\ell yz^{-1}y^{-1}$. The differentials are given by

$$\partial_1(e_x) = x - 1, \quad \partial_1(e_y) = y - 1 \quad \text{and} \quad \partial(e_z) = z - 1; \quad \text{and}$$

$$\partial_2(r) = (z^f y - 1)e_x + (z^f - x)e_y + (\Sigma_{i=0}^{f-1} z^i)e_z, \quad \partial_2(s) = \nu e_z,$$

$$\partial_2(t) = (z - 1)e_x + (1 + zx)e_z \quad \text{and} \quad \partial_2(u) = (z^\ell - 1)e_y + (\Sigma_{j=0}^{\ell-1} z^j - y)e_z.$$

We may choose a homomorphism $\partial_3 : C_3 \to C_2$ with domain C_3 a free $\mathbb{Z}[G]$-module and image $\mathrm{Ker}(\partial_2)$, which extends the resolution one step to the left. (We may assume that C_3 is finitely generated, since $\mathbb{Z}[G]$ is noetherian.) It is clear from the Fox-Lyndon partial resolution that $\dim_{\mathbb{F}_2} \mathrm{Ker}(\mathbb{F}_2 \otimes_{\mathbb{Z}[G]} \partial_2) = \beta_1(G; \mathbb{F}_2) + 1$. We shall show that $\mathbb{F}_2 \otimes_{\mathbb{Z}[G]} \partial_3 = 0$, and so $\beta_2(G; \mathbb{F}_2) = \beta_1(G; \mathbb{F}_2) + 1$.

Let $\varepsilon_2 : \mathbb{Z}[G] \to \mathbb{F}_2$ be the mod (2) reduction of ε. Since $\mathrm{Im}(\partial_3) = \mathrm{Ker}(\partial_2)$, it shall suffice to show that if

$$\partial_2(ar + bs + ct + du) = 0,$$

for some $a, b, c, d \in \mathbb{Z}[G]$ then $\varepsilon_2(a) = \varepsilon_2(b) = \varepsilon_2(c) = \varepsilon_2(d) = 0$.

The coefficients a, b, c, d must satisfy the three equations

$$a(z^f y - 1) + c(z - 1) = 0,$$

$$a(z^f - x) + d(z^\ell - 1) = 0$$

and

$$a(\Sigma_{i=0}^{f-1} z^i) + b\nu + c(zx + 1) + d(\Sigma_{j=0}^{\ell-1} z^j - y) = 0.$$

Multiplying the first of these equations by ν gives $af\nu(y - 1) = 0$. Hence $a\nu = 0$ and so $a = A(z-1)$, for some $A \in \mathbb{Z}[G]$ not involving z. The first equation becomes

$$A(z - 1)(z^f y - 1) + c(z - 1) = [A(yz^{fm}(\Sigma_{j=0}^{m-1} z^j) - 1) + c](z - 1) = 0,$$

and so $c = -A(yz^{fm}(\Sigma_{j=0}^{m-1}z^j) - 1) + C\nu$, for some $C \in \mathbb{Z}[G]$ not involving z. Similarly, the second equation becomes

$$A(z-1)(z^f - x) + d(z^\ell - 1) = A(zx + z^f)(z^{m\ell} - 1) + d(z^\ell - 1) = 0,$$

and so $d = -A(zx + z^f)(\Sigma_{j=0}^{m-1}z^{j\ell}) + D\nu$, for some $D \in \mathbb{Z}[G]$ not involving z. At this point it is already clear that $\varepsilon_2(a) = \varepsilon_2(c) = \varepsilon_2(d) = 0$.

Multiplying the third equation by ν gives

$$kb\nu + c\nu(x+1) + d\nu(\ell - y) = 0.$$

When this equation is rearranged and written out in full, it becomes

$$kb\nu = (A(ym-1) - Ck)(x+1)\nu + (A(x+1)m - Dk)(\ell - y)\nu.$$

Since $yx = z^{-f}xy = xzy = xyz^{fm}$ we have $yx\nu = xy\nu$ and so this simplifies to

$$kb\nu = (A(m\ell - 1)(x+1) - kC(x+1) - kD(\ell - y))\nu.$$

Write $b = b_1 + B(z-1)$, where b_1 does not involve z. Then $b\nu = b_1\nu$. Since the terms b_1, A, C and D do not involve z, and since $m\ell - 1 = wk$, we get

$$kb_1 = k(Aw(x+1) - C(x+1) - D(\ell - y)).$$

We may solve for b_1, and so

$$b = b_1 + B(z-1) = wA(x-1) + B(z-1) - C(x+1) - D(\ell - y).$$

Hence $\varepsilon_2(b) = 0$ also, so $\mathbb{F}_2 \otimes_{\mathbb{Z}[G]} \partial_3 = 0$ and thus $\beta_2(G;\mathbb{F}_2) = \beta_1(G;\mathbb{F}_2) + 1$. □

LEMMA B.13. *Let G be a finitely generated nilpotent group, and let T be its torsion subgroup. If $G/T \cong \mathbb{Z}^2$ and the Sylow p-subgroup of T is abelian and non-trivial then $\beta_2(G;\mathbb{F}_p) > \beta_1(G;\mathbb{F}_p)$.*

PROOF. Let N be the product of all the Sylow p'-subgroups of T with $p' \neq p$, and let A be the image of T in $\overline{G} = G/N$. Then $\beta_i(\overline{G};\mathbb{F}_p) = \beta_i(G;\mathbb{F}_p)$, for all i. The Sylow p-subgroup of T projects isomorphically onto A, and $\overline{G}/A \cong \mathbb{Z}^2$. If $\beta_2(G;\mathbb{F}_p) \leqslant \beta_1(G;\mathbb{F}_p)$ then $\dim_{\mathbb{F}_p} H^2(A;\mathbb{F}_p)^{G/A} \leqslant 1$, by Lemma B9. Hence A is cyclic, by Lemmas B2 and B3. If $p = 2$ the result follows from Lemma B12 while if p is odd it follows from Lemma B11 since the automorphism group of a cyclic group of odd p-power order is cyclic. □

For the next result we need an analogue of Lemma B10.

LEMMA B.14. *Let $K \cong T \rtimes \mathbb{Z}^2$, where T is a finite p-group, and $T \not\leqslant K'$. Then $H^2(K;\mathbb{F}_p)$ is canonically subsplit.*

PROOF. Let $\alpha : K \to \mathbb{Z}^2$ be the canonical epimorphism. Since α splits, $H^2(\alpha;\mathbb{F}_p)$ is a monomorphism. The other hypotheses imply $Ext(K^{ab},\mathbb{F}_p) \neq 0$. Hence $Ext(K^{ab},\mathbb{F}_p) \oplus \mathrm{Im}(H^2(\alpha;\mathbb{F}_p))$ is a subspace of $H^2(K;\mathbb{F}_p)$ with the desired properties. □

If G is a homologically balanced, metabelian nilpotent group then either $G \cong \mathbb{Z}^3$ or $\beta_1(G;\mathbb{Q}) \leqslant 2$ and $h(G) \leqslant 4$. In the latter case the torsion-free quotient G/T is either free abelian of rank $\leqslant 2$, or is the $\mathbb{N}il^3$-group Γ_q, for some $q \geqslant 1$ [**Hi22**, Corollary 8 and Theorems 10 and 15].

THEOREM B.15. *Let G be a homologically balanced nilpotent group with $\beta_1(G;\mathbb{Q}) = 2$, and let T be its torsion subgroup. Then*

(1) *if $h(G) = 2$ and T is abelian then $G \cong \mathbb{Z}^2$;*
(2) *if $h(G) = 3$ and the outer action : $G \to Out(T)$ determined by conjugation in G factors through $\mathbb{Z}^2$ then $G \cong \Gamma_q$, for some $q \geqslant 1$;*
(3) *if $h(G) = 4$ and G has an abelian normal subgroup A with $G/AT \cong \mathbb{Z}^2$ then $G \cong \Omega$.*

PROOF. Let $K = I(G)$. Then $TG' \leqslant K$, $G/K \cong \mathbb{Z}^2$, and K/G' is (finite) cyclic, since $\beta_1(G;\mathbb{F}_p) \leqslant 3$ for all primes p.

If $h(G) = 2$ then $K = T$ and $G/T \cong \mathbb{Z}^2$. Hence if T is abelian then $T = 1$, by Lemma B13, and so $G \cong \mathbb{Z}^2$.

If $h(G) = 3$ and the outer action : $G \to Out(T)$ determined by conjugation in G factors through $\mathbb{Z}^2$ then $K \cong T \times \mathbb{Z}$. Thus if $T \neq 1$ then $\beta_2(G;\mathbb{F}_p) > \beta_1(G;\mathbb{F}_p)$, for any prime p dividing $|T|$, by Lemmas B9 and B10. This contradicts the hypothesis that G has a balanced presentation.

If $h(G) = 4$ and G has an abelian normal subgroup A such that $G/AT \cong \mathbb{Z}^2$ then $K = AT$, and $\overline{A} = A/A \cap T \cong \mathbb{Z}^2$. Since K is nilpotent, the action of the finite group $T/A \cap T$ on $\overline{A}$ is trivial. Hence $K/A \cap T \cong \overline{A} \times (T/A \cap T)$, and so $K \cong T \rtimes \mathbb{Z}^2$. Moreover, if $T \neq 1$ then $T \nleqslant K'$. Lemmas B9 and B14 (together with Lemmas B2 and B3) then give a similar contradiction.

In parts (2) and (3) the group G is torsion-free, and so must be one of the known examples given above. □

Imposing a stronger constraint gives a clearer statement.

COROLLARY. *If G is a homologically balanced nilpotent group with an abelian normal subgroup A such that $G/A \cong \mathbb{Z}^2$ then $G \cong \mathbb{Z}^2$, Γ_q (for $q \geqslant 1$) or Ω.* □

Note that the second hypotheses in parts (2) and (3) of the theorem are not by themselves equivalent to assuming that G is metabelian, while the hypothesis in the corollary is somewhat stronger. (On the other hand, it includes all 2-generated metabelian nilpotent groups G with $h(G) > 1$.)

The above work leaves open the following questions, for G a nilpotent group with torsion subgroup T and a balanced presentation.

(1) If $h(G) = 1$ is $H_2(T) = 0$?
(2) If $h(G) > 1$ is $T = 1$?

Appendix C Some Questions

This list includes questions from [**BB22, IM20**] and [**Liv05**]. We have included several questions specifically about smooth embeddings, but many of the other questions have obvious smooth analogues.

As in the text, M is a closed orientable 3-manifold, and if $j : M \to S^4$ is a locally flat embedding then the complementary regions X and Y are labelled so that $\chi(X) \leqslant \chi(Y)$. The fundamental groups are $\pi = \pi_1 M$, $\pi_X = \pi_1 X$ and $\pi_Y = \pi_1 Y$. We shall abbreviate "embeds in S^4" to "embeds". If a link L is "bipartedly slice" then the 3-manifold $M(L)$ obtained by 0-framed surgery on L embeds.

(1) Is there an effective way of deciding whether a framed link in S^3 is equivalent under Kirby moves to a 0-framed bipartedly slice link?

(2) If each complementary region for an embedding j may be obtained from the 4-ball by adding 1- and 2-handles, must $j = j_L$ for some 0-framed bipartedly trivial link L?

(3) [**BB22**] Is there a 3-manifold M which embeds smoothly, but which has no smooth embedding deriving from a 0-framed link presentation?

(4) (Attributed to M. H. Freedman in [**BB22**].) If $M\#(S^2 \times S^1)$ embeds, does M embed?

Freedman and Krushkal have found an invariant which detects non-embeddability for certain 3-manifolds M with $\tau_M = H_1(M)$ and ℓ_M hyperbolic [**FK25**].

(5) How is the invariant λ_3 defined in [**FK25**] related to other more familiar invariants of 3-manifolds?

Most of the algebraic arguments obstructing embeddings of a given 3-manifold into S^4 that we know of also obstruct Poincaré embeddings in homology 4-spheres.

(6) Is there a 3-manifold M which has a Poincaré embedding in a homology 4-sphere but which does not embed?

There are homology spheres which do not embed smoothly in any smooth homotopy 4-sphere, but do embed smoothly in some smooth homology 4-sphere [**Mc22**].

(7) Let $G = Q(8a, b, c)$, where a, b, c are odd and relatively prime. For which values of a, b, c is there a finite PD_3-complexes P with $\pi_1 P \cong G$ which has a Poincaré embedding in a homology 4-sphere?

If $G = Q(8a)$ then the 3-manifold S^3/G fibres over $\mathbb{RP}^2$, and so bounds a disc bundle over $\mathbb{RP}^2$. However $S^3/Q(8a)$ does not embed in a homology 4-sphere if $a > 1$, by Corollary 4.4.2.

(8) Is there a homology 3-sphere with a free G-action, where $G = Q(8a, b, c)$ with a, b, c odd and relatively prime and $a > b > c \geqslant 1$?

Questions 7 and 8 are prompted by the final section of Chapter 6.

(9) If j is an embedding of $T_g \times S^1$ such that $j_{X*} : \pi \to \pi_X$ is an isomorphism is j s-concordant to the standard embedding?

The standard embedding is as the boundary of a regular neighbourhood of the unknotted embedding $T_g \subset S^3 \subset S^4$. (See Chapter 7.)

If M embeds and has non-trivial homology then $\pi_X \neq 1$, and we may use 2-knot surgery to construct infinitely many embeddings of M, distinguished by the groups π_X. (See Chapter 2.) On the other hand, if $M = S^3$ then $\pi_X = \pi_Y = 1$, and all embeddings of S^3 are essentially equivalent, by the Schoenflies Theorem. The question remains open for homology 3-spheres.

(10) Does every homology 3-sphere bound a non-simply connected acyclic 4-manifold? have an embedding with one complementary region not 1-connected? have an embedding with neither complementary region 1-connected?

Similarly, if $M = S^2 \times S^1$ then $\pi_Y = 1$, by Aitchison's Theorem (Theorem 7.2).

(11) Does every homology handle other than $S^2 \times S^1$ have embeddings with neither complementary region 1-connected?

There are uniform non-embedding results for Seifert fibred homology handles, but the known constructions (see Chapter 3 and [**Do15, IM20**]) apply only under restrictions on cone points of even order.

If the exceptional fibres of a Seifert manifold $M(k; S)$ occur in pairs $(\alpha, \pm\beta)$ we say that the Seifert data S is skew-symmetric. (See Chapter 3.)

(12) Are there any restrictions related to 2-torsion in the cone point orders of the base orbifolds of Seifert manifolds which embed?
(13) Let ℓ be a hyperbolic linking pairing with homogeneous 2-primary summand $\ell_{(2)}$ (as defined in Appendix A). Is ℓ realized by some Seifert manifold $M(0; S)$ which embeds?
(14) Let $M = M(0; S)$ where S is skew-symmetric, all even cone point orders are divisible by the same power of 2 and ℓ_M is hyperbolic. Does M embed? In particular, does $M(0; (2, 1), (2, -1), (10, 3), (10, -3))$ embed?
(15) If a Seifert fibred homology sphere embeds smoothly must it have at most three exceptional fibres?

Note that (13) and (14) are not the same question!

The work of Donald and of Issa and McCoy on smooth embeddings relies on gauge theory, through Donaldson's Diagonalization Theorem. We shall paraphrase two of their conjectures in terms of TOP locally flat embeddings.

(16) [**IM20**, Conjecture 1.8]. If a Seifert manifold $M(0; S)$ with $\varepsilon(M) > 0$ embeds, is it the result of a (possibly empty) sequence of expansions from some $M(0; S', (1, -1))$ with S' strict Seifert data which also embeds.

(17) [**IM20**, Conjecture 1.9]. If $M = M(0; S', (1, -e))$ embeds and $\varepsilon(M) > 0$ then is $k - 1 \leqslant 2e \leqslant k + 1$?

(18) Which connected sums of lens spaces embed?

Graph manifolds have natural parametrizations in terms of weighted plumbing graphs. See [**Neu81**] and [**Orl**, Chapter 2].

(19) What can be said about TOP locally flat embeddings of graph manifolds?

(20) Do the arguments of Donald and Issa and McCoy extend to the study of smooth embeddings of graph manifolds?

In the remaining questions the emphasis is on the complementary regions.

(21) Is there a 3-manifold with at least two abelian embeddings which are inequivalent?

(22) Does the $\mathbb{S}ol^3$-manifold $M_{2,4}$ have an abelian embedding?

Whether there is an abelian embedding has been decided for all other 3-manifolds with virtually solvable fundamental groups. (See Chapter 8.)

An embedding j is bi-epic if the associated homomorphisms from π to π_X and π_Y are each onto. If $\mathcal{P}$ is an adjective describing a class of groups then j is $\mathcal{P}$ if π_X and π_Y are each $\mathcal{P}$. A group is restrained if it has no non-cyclic free subgroup.

(23) If j is a restrained bi-epic embedding is $\chi(X) \geqslant 0$, and is X aspherical if $\chi(X) = 0$?

This is so if $j = j_L$ for some 0-framed bipartedly trivial link, by Theorem 9.7.

(24) If an embedding j is bi-epic and virtually solvable, elementary amenable or restrained, is $\beta \leqslant 6$?

This is really an algebraic question about the rank of G^{ab} for a homologically balanced group G in one of these (successively larger) classes. It is true for bi-epic embeddings which are virtually polycyclic, by Theorem 9.4. One could ask the same question for groups which are virtually solvable and of finite Hirsch length or for which $\beta_1^{(2)}(G) = 0$.

(25) Is there an embedding with X aspherical and $c.d.\pi_X = 3$?

If X is aspherical it cannot be a PD_3-complex, since $H_3(X) = 0$, and so π_X cannot be a PD_3-group. See also [**DH25**], which considers aspherical 4-manifolds with elementary amenable fundamental group and non-empty boundary.

(26) What are the possible values of $\chi(X)$ for embeddings of the Seifert manifold $M(k; S)$?

(27) Is there a 3-manifold which embeds, but has no embedding with $\chi(X) = \chi(Y) = 1$ (if β is even) or $\chi(X) = 0$ and $\chi(Y) = 2$ (if β is odd)? In particular, does the Seifert manifold $M(-3; (1, 6))$ have an embedding with $\chi(X) = \chi(Y) = 1$?

The S^1-bundle space $M(-3; (1, 6))$ bounds embeddings of $\#^3\mathbb{RP}^2 = T\#\mathbb{RP}^2$, which all have $\chi(X) = -1$ and $\chi(Y) = 3$. (In this case $\beta = 2$.)

(28) [**Liv05**] If G is the fundamental group of a homology 4-sphere, is $G = \pi_1 W$, where $W \subset S^4$ is a codimension-0 submanifold with contractible complement?

The converse is clear. If N is a regular open neighbourhood of a contractible submanifold of S^4 then $W = S^4 \setminus N$ is acyclic, its double DW is a homology 4-sphere, and $\pi_1 DW \cong \pi_1 W$.

There is a smooth embedding j such that one of the groups π_X or π_Y has an unsolvable word problem [**DR93**].

(29) [**BB22**] If M embeds in S^4, does it have an embedding such that π_X and π_Y each have solvable word problem?

Finally

(30) What can be said about complementary regions of hypersurfaces with more than one component?

Added in proof. Question 21 has been settled [**Hi25**]. Replacing one component of the Borromean rings Bo by a hyperbolic knot with Alexander polynomial 1 (such as the Kinoshita-Teresaka knot 11_{42n}) gives a a link L for which $M(L)$ is an example, with $\beta = 3$. (See the figure.) Variations on this construction give examples with $\beta = 2, 4$ or 6. (One example with $\beta = 6$ has at least five distinct abelian embeddings.) Uniqueness of JSJ decompositions is used to distinguish abelian embeddings associated to different partitions of the link.

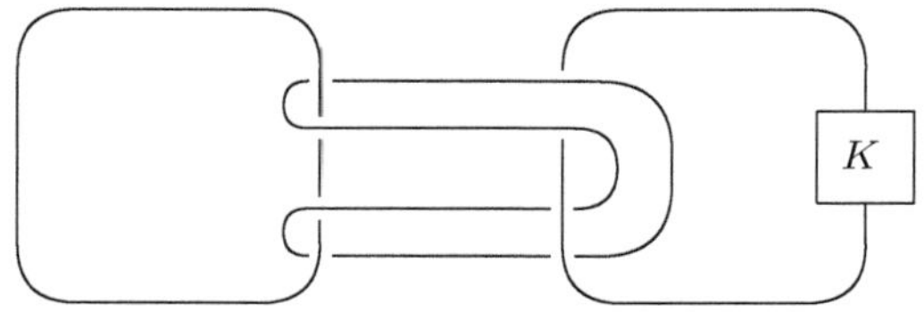

In Section 8.6 we made some brief observations about using surgery to classify abelian embeddings of 3-manifolds M with $H_1(M) \cong C_n^2$ in S^4, for the cases $n \leqslant 4$. We shall remove the latter restriction here. Let W be a complementary region of an embedding of a rational homology sphere M in S^4, such that $\pi_W = \pi_1 W \cong C_n$, for some $n \geqslant 1$. Then $W \simeq P_n = S^1 \cup_n e^2$, by Lemma 8.4, and homotopy equivalences between such pseudoprojective planes are simple [**DS73**]. Exactness of the surgery sequence implies that $L_1^s(\mathbb{Z}[\pi_W])$ acts transitively on the structure set $\mathcal{S}_{TOP}(W, M)$, since $H^2(W, M; \mathbb{F}_2) = H_2(W; \mathbb{F}_2) = 0$, and so normal invariants are detected by the signature. If moreover n is odd then $L_1^s(\mathbb{Z}[\pi_W]) = 0$ [**Bak75**], and the complementary regions of an abelian embedding of M are determined up to homeomorphism by the homotopy types of the inclusions of their boundaries. (If n is even then $L_1^s(\mathbb{Z}[\pi_W])$ is a finite 2-group [**Wa76**].)

Bibliography

Books

[BKKPR] Behrends, S., Kalmár, B., Kim, M. H., Powell, M. and Ray, A. *The Disc Embedding Theorem*, Oxford University Press, Oxford (2021).

[Bie] Bieri, R. *Homological Dimension of Discrete Groups*, Queen Mary College Mathematics Notes, University of London, London (1976).

[Bro] Brown, K. S. *Cohomology of Groups*, Graduate Texts in Mathematics 87, Springer-Verlag, Berlin (1982).

[Coh] Cohen, M. M. *A Course in Simple-Homotopy Theory*, Graduate Texts in Mathematics 10, Springer-Verlag, Berlin (1973).

[DD] Dicks, W. and Dunwoody, M. J. *Groups Acting on Graphs*, Cambridge Studies in Advanced Mathematics 17, Cambridge University Press, Cambridge (1989).

[FQ] Freedman, M. H. and Quinn, F. *Topology of 4-Manifolds*, Princeton University Press, Princeton (1990).

[FNOP] Friedl, S., Nagel, M., Orson, P. and Powell, M. *The Foundations of Four-Manifold Theory in the Topological Category*,New York J. Math. Monograph 6 , New York (2025).

[GL] Guirardel, V. and Levitt, G. *JSJ Decompositions of Groups*, Astérisque 395, Société Mathématique de France, Paris (2017).

[GM] Gillou, L. and Marin, A. (editors) *Á la recherche de la topologie perdue*, Progress in Mathematics vol. 62, Birkhäuser Verlag, Basel (1986).

[GS] Gompf, R. and Stipsicz, A. *4-Manifolds and Kirby Calculus*, Graduate Studies in Mathematics 20, American Mathematical Society, Providence (1999).

[Hem] Hempel, J. *3-Manifolds*, Annals of Mathematics Study 86, Princeton University Press, Princeton (1976).

[AIL] Hillman, J. A. *Algebraic Invariants of Links*, second edition, Series on Knots and Everything, vol. 52, World Scientific Publishing, Singapore (2012).

[FMGK] Hillman, J. A. *Four-Manifolds, Geometries and Knots*, Geometry and Topology Monographs, vol. 5, Geometry and Topology Publications, Warwick (2002). (Latest revision 2021.)

[JN] Jankins, M. and Neumann, W. D. *Lectures on Seifert Manifolds*, Brandeis Lecture Notes 2, Brandeis University, Waltham (1983).

[HAMS] Hog-Angeloni, C., Metzler, W. and Sieradski, A. J. *Two-Dimensional Homotopy and Combinatorial Group Theory*, London Lecture Notes Series 197, Cambridge University Press, Cambridge (1993).

[John] Johnson, F. E. A. *Stable Modules and the D(2)-Problem*, London Lecture Notes Series 301, Cambridge University Press, Cambridge (2003).

[KL] Kreck, M. and Lück, W. *The Novikov Conjecture – Geometry and Algebra*, Oberwolfach Seminars vol. 33, Birkhäuser Verlag, Basel (2005).

[Lück] Lück, W. *L^2-Invariants: Theory and Applications to Geometry and K-Theory*, Ergebnisse der Mathematik und ihrer Grenzgebiete 3 Folge, Bd. 44, Springer-Verlag, Berlin (2002).

[Neu] Neumann, W. D. *Equivariant Witt Rings*, Bonner Mathematische Schriften 100, Mathematisches Institut der Universität, Bonn(1977).

[Orl] Orlik, P. *Seifert Manifolds*, Lecture Notes in Mathematics 291, Springer-Verlag, Berlin (1972).

[Ran] Ranicki, A. A. *Algebraic and Geometric Surgery*, Oxford Mathematical Monographs, Oxford University Press, Oxford (2002).

[Rob] Robinson, D. J. S. *A Course in the Theory of Groups*, Graduate Texts in Mathematics 80, Springer-Verlag, Berlin (1982).

[Rol] Rolfsen, D. *Knots and Links*, Mathematics Lecture Series 7, Publish or Perish, Berkeley (1976).

[Ser] Serre, J.-P. *Galois Cohomology*, Springer Monographs in Mathematics, Springer-Verlag, Berlin (1997).

[Span] Spanier, E. H. *Algebraic Topology*, McGraw-Hill Series in Higher Mathematics, McGraw-Hill, New York (1966).

[vRan] von Randow, R. *Zur Topologie von drei-dimensionalen Baumannigfaltigkeiten*, Bonner Mathematische Schriften 14, Mathematisches Institut der Universität, Bonn (1962).

[Wall] Wall, C. T. C. *Surgery on Compact Manifolds*, second edition, edited and with a foreword by A. A. Ranicki, Mathematical Surveys and Monographs 69, American Mathematical Society, Providence (1999).

Journal Articles

[AJOT13] Abe, T., Jong, I., Omae, Y. and Takeuchi, M. Annulus twist and diffeomorphic 4-manifolds, Math. Proc. Cambridge Phil. Soc. 155 (2013), 219–235.

[AMP22] Aceto, P., McCoy, D. and Park, J. H. Definite fillings of lens spaces, J. Diff. Geom. (Also arXiv: 2208.02586 [math.GT].)

[AMP24] Aceto, P., McCoy, D. and Park, J. H. A survey on embeddings of 3-manifolds in definite 4-manifolds, *Proceedings of MSJ-KMS Joint Meeting, Sendai, 2023.* (Also arXiv: 2407.03692 [math.GT].)

[Al24] Alexander, J. W. An example of a simply connected surface bounding a region which is not simply connected, Nat. Acad. Proc. 10 (1924), 8–10

[Ar81] Artamanov, V. A. Projective non-free modules over group rings of solvable groups, Mat. Sbornik 116 (1981), 232–244. [Russian]

[AB68] Atiyah, M. F. and Bott, R. A Lefschetz fixed point theorem for elliptic complexes. II. Applications, Ann. Math. 88 (1968), 451–491.

[AS68] Atiyah, M. F. and Singer, I. M. The index theorem for elliptic operators: III, Ann. Math. 87 (1968), 546–604.

[Bak75] Bak, A. Odd dimension surgery groups of odd torsion groups vanish, Topology 14 (1975), 367–374.

[Ba76] Bass, H. Euler characteristics and characters of discrete groups, Invent. Math. 35 (1976), 155–196.

[BH17] Bauval, A. and Hayat, C. L'anneau de cohomologie des variétés de Seifert non-orientables, Osaka J. Math. 54 (2017), 157–195.

[BPW19] Bellettini, G., Paolini, M. and Wang, Y. S., A complete invariant for closed surfaces in the 3-sphere, J. Knot Theory Ramif. 30 (2021), 2150044.

[Bo86] Boyer, S. Simply-connected 4-manifolds with given boundary, Trans. Amer. Math. Soc. 298 (1986), 331–357.

[Bry06] Bryant, R. L. Remarks on the geometry of almost complex 6-manifolds, Asian J. Math. 10 (2006), 561–606.

[BD04] Bryden, J. and Deloup, F. A linking form conjecture for 3-manifolds, in *Advances in Topological Quantum Field Theory*, NATO Sci. Ser. II Math. Phys. Chem., 179, Kluwer, Dordrecht (2004), 253–265.

[BZ03] Bryden, J. and Zvengrowski, P. The cohomology ring of the orientable Seifert manifolds, II, Topology Appl. 127 (2003), 213–257.

[BB22] Budney, R. and Burton, B. A. Embedding of 3-manifolds in S^4 from the point of view of the 11-tetrahedron census, Experimental Math. 31 (2022), 988–1013.

[CG83] Casson, A. and Gordon, C. McA. A loop theorem for duality spaces and fibred ribbon knots, Invent. Math. 74 (1983), 119–137.

[CH90] Cochran, T. D. and Habegger, N. On the homotopy theory of simply-connected four-manifolds, Topology 29 (1990), 419–440.

[CT14] Cochran, T. D. and Tanner, D. Homology cobordism and Seifert fibred 3-manifolds, Proc. Amer. Math. Soc. 142 (2014), 4015–4024.

[Co71] Connelly, R. A new proof of Brown's collaring theorem, Proc. Amer. Math. Soc. 27 (1971), 180–182.

[Cr88] Craggs, R. On the algebra of handle operations in 4-manifolds, Top. Appl. 30 (1988), 237–252.

[CH98] Crisp, J. S. and Hillman, J. A. Embedding Seifert fibred 3-manifolds and $\mathbb{S}ol^3$-manifolds in 4-space, Proc. London Math. Soc. 76 (1998), 685–710.

[Da94] Daverman, R. J. Fundamental group isomorphisms between compact 4-manifolds and their boundaries, in *Low-dimensional Topology (Knoxville, TN, 1992)*, edited by K. Johannson, Conference Proceeedings and Lecture Notes in Geometry and Topology III, International Press, Cambridge (1994), 31–34.

[DH25] Davis, J. F. and Hillman, J. A. Aspherical 4-manifolds with boundary and elementary amenable groups, arXiv: 2501.12512 [math.GT].

[De05] Deloup, F. Monoide des enlacements at facteurs orthogonaux, Alg. Geom. Top. 5 (2005), 419–442.

[DH17] Doig, M. I. and Horn, P. D. On the intersection ring of graph manifolds, Trans. Amer. Math. Soc. 369 (2017), 1185–1203.

[Do15] Donald, A. Embedding Seifert manifolds in S^4, Trans. Amer. Math. Soc. 367 (2015), 559–595.

[Do87] Donaldson, S. The orientation of Yang-Mills moduli spaces and 4-manifold topology, J. Diff. Geom. 26 (1987), 397–428.

[DR93] Dranishnikov, A. and Repovš, D. Embeddings up to homotopy type in Euclidean Space, Bull. Austral. Math. Soc. 47 (1993), 145–148.

[Dw75] Dwyer, W. G. Homology, Massey products and maps between groups, J. Pure Appl. Alg. 6 (1975), 177–190.

[DS73] Dyer, M. and Sieradski, A. Trees of homotopy types of two-dimensional CW-complexes, Comment. Math. Helv. 48 (1973), 31–44.

[Ec01] Eckmann, B. Idempotents in a complex group algebra, projective modules, and the von Neumann algebra, Archiv Math. (Basel) 76 (2001), 241–249.

[Ed05] Edmonds, A. L. Homology lens spaces in topological 4-manifolds, Illinois J. Math. 49 (2005), 827–837.

[EE82] Edmonds, A. L. and Ewing, J. H. Remarks on the cobordism group of surface diffeomorphisms, Math. Ann. 259 (1982), 497–504.

[EL96] Edmonds, A. L. and Livingston, C. Embedding punctured lens spaces in four-manifolds, Comment. Math. Helv. 71 (1996), 169–191.

[Ed06] Edwards, T. Generalized Swan modules and the $D(2)$ problem, Alg. Geom. Top. 6 (2006), 71–89.

[Ep61] Epstein, D. B. A. Factorization of 3-manifolds, Comment. Math. Helv. 36 (1961), 91–102.

[Ep65] Epstein, D. B. A. Embedding punctured manifolds, Proc. Amer. Math. Soc. 16 (1965), 175–176.

[FJ88] Farrell, F. T. and Jones, L. P. The surgery L-groups of poly-(finite or cyclic) groups, Invent. Math. 91 (1988), 559–586.

[FS84] Fintushel, R. and Stern, R. $SO(3)$-connections and the topology of 4-manifolds, J. Diff. Geom. 30 (1984), 523–539.

[Fr82] Freedman, M. H. The topology of four-dimensional manifolds, J. Diff. Geom. 17 (1982), 357–453.

[Fr24] Freedman, M. H. Enhanced Hantzsche Theorem, arXiv: 2409. 09983 [math.GT].

[FHT97] Freedman, M., Hain, R. and Teichner, P. Betti number estimates for nilpotent groups, in *Fields Medallists Lectures*, edited by M. Atiyah and D. Iagolnitzer, WSP Series in 20th Century Mathematics 5, World Scientific Publications, Singapore (1997), 413–434.

[FK25] Freedman, M. H. and Krushkal, V. A triple torsion linking form for a 3-manifold in S^4, arXiv: 2506.11941 [math.GT].

[FT95] Freedman, M. H. and Teichner, P. 4-Manifold topology I: subexponential groups, Invent. Math. 122 (1995), 509–529.

[Ga87] Gabai, D. Foliations and the topology of 3-manifolds: III, J. Diff. Geom. 26 (1987), 479–536.

[GL83] Gilmer, P. M. and Livingston, C. On embedding 3-manifolds in 4-space, Topollogy 22 (1983), 241–252.

[Go86] Gordon, C. McA. On the G-signature theorem in dimension four, in [**GM**], 159–180.

[GJ11] Greene, J. and Jabuka, S. The slice-ribbon conjecture for 3-stranded pretzel knots, Amer. J. Math. 133 (2011), 550–580.

[GK92] Grigorchuk, R. I. and Kurchanov, P. F. On quadratic equations in groups, in *Proceedings of the International Conference on Algebra. Dedicated to the Memory of A. I. Mal'cev*, edited by L. A. Bokut', A. I. Mal'cev and A. I. Kostrikin, CONM 131, American Mathematical Society, Providence (1992), 159–171.

[Ha38] Hantzsche, W. Einlagerung von Mannigfaltigkeiten in euklidische Räume, Math. Z. 43 (1938), 38–58.

[HW85] Hausmann, J.-C. and Weinberger, S. Caractéristiques d'Euler et groupes fondamentaux des variétés en dimension 4, Comment. Math. Helv. 60 (1985), 139–144.

[HNO'B96] Havas, G., Newman, M. F. and O'Brien, E. A. Groups of deficiency zero, in *Geometric and Computational Perspectives on Infinite Groups*, DIMACS Series in Discrete Mathematics and Theoretical Computer Science 25, Amer. Math. Soc. (1996), 53–67.

[Hg51] Higman, G. A finitely generated infinite simple group, J. London Math. Soc. 26 (1951), 61–64.

[Hi87] Hillman, J. A. The kernel of integral cup product, J. Austral. Math. Soc. 43 (1987), 10–15.

[Hi96] Hillman, J. A. Embedding homology equivalent 3-manifolds in 4-space, Math. Z. 223 (1996), 473–481.

[Hi09] Hillman, J. A. Embedding 3-manifolds with circle actions, Proc. Amer. Math. Soc. 137 (2009), 4287–4294.

[Hi11] Hillman, J. A. The linking pairings of orientable Seifert manifolds, Top. Appl. 158 (2011), 468–478.

[Hi17] Hillman, J. A. Complements of connected hypersurfaces in S^4, Special Volume in Memory of Tim Cochran, J. Knot Theory Ramif. 26 (2017), 1740014.

[Hi20a] Hillman, J. A. 3-manifolds with abelian embeddings in S^4, J. Knot Theory Ramif. 29 (2020), 2050001.

[Hi20b] Hillman, J. A. 3-manifolds with nilpotent embeddings in S^4, J. Knot Theory Ramif. 29 (2020), 2050094.

[Hi22] Hillman, J. A. Nilpotent groups with balanced presentations, J. Group Theory 25 (2022), 713–726.

[Hi25] Hillman, J. A. 3-manifolds with more than one abelian embedding, arXiv: 2506.01186 [math.GT].

[Hu90] Huck, G. Embeddings of acyclic 2-complexes in S^4 with contractible complement, in *Topology and Combinatorial Group Theory*, Lecture Notes in Mathematics 1440, Springer-Verlag, Berlin (1990), 122–129.

[IM20] Issa, A. and McCoy, D. Smoothly embedding Seifert fibered spaces in S^4, Trans. Amer. Math. Soc. 373 (2020), 4933–4974.

[IZ18] Ivanov, S. O. and Zaikovskii, A. A. Mod-2 (co)homology of an abelian group, J. Math. Sciences 252 (2021), 794–803.

[Ka83] Kanenobu, T. Groups of higher dimensional satellite knots, J. Pure Appl. Alg. 28 (1983), 179–188.

[Ka79] Kaplan, S. J. Constructing framed 4-manifolds with given almost-framed boundaries, Trans. Amer. Math. Soc. 254 (1979), 237–263.

[Ka77] Kawauchi, A. On quadratic forms of 3-manifolds, Invent. Math. 43 (1977), 177–198.

[Ka79] Kawauchi, A. On n-manifolds whose punctured manifolds are imbeddable in $(n + 1)$-spheres and spherical manifolds, Hiroshima Math. J. 9 (1979), 47–57.

[Ka88] Kawauchi, A. The imbedding problem of 3-manifolds into 4-manifolds, Osaka J. Math. 25 (1988), 171–183.

[KK80] Kawauchi, A. and Kojima, S. Algebraic classification of linking pairings on 3-manifolds, Math. Ann. 253 (1980), 29–42.

[Kir] Kirby, R. C. Problems in low-dimensional topology, in *Geometric Topology*, edited by W. H. Kazez, Part 2, American Mathematical Society, Providence (1997), 35–473.

[KT02] Kirby, R. C. and Taylor, L. R. A survey of 4-manifolds through the eyes of surgery, in *Surveys on Surgery Theory*, vol. 2, edited by A.A. Ranicki, Ann. Math. Study 149, Princeton University Press, Princeton NJ (2002), 387–421.

[Law84] Lawson, T. Detecting the standard embedding of RP^2 in S^4, Math. Ann. 267 (1984), 439–448.

[Lev83] Levine, J. P. Doubly sliced knots and doubled disc knots, Michigan Math. J. 30 (1983), 249–256.

[Lic03] Lickorish, W. B. R. Knotted contractible 4-manifolds in the 4-sphere, Pacific. J. Math. 208 (2003), 283–290.

[Lic04] Lickorish, W. B. R. Splittings of S^4, Bol. Soc. Mat. Mexicana 10 (2004), Special Issue, 305–312.

[Li07] Lisca, P. Sums of lens spaces bounding rational balls, Alg. Geom. Top. 7 (2007), 2141–2164.

[Lit84] Litherland, R. A. Cobordism of satellite knots, CONM 35, Amer. Math. Soc. (1984), 327–362.

[Liv81] Livingston, C. Homology cobordisms of 3-manifolds, Pac. J. Math. 94 (1981), 193–206.

[Liv03] Livingston, C. Observations on Lickorish knotting of contractible 4-manifolds, Pac. J. Math. 209 (2003), 319–323.

[Liv05] Livingston, C. Four-manifolds with large negative deficiency, Math. Proc. Cambridge Phil. Soc. 138 (2005), 107–115.

[Lub83] Lubotzky, A. Group presentation, p-adic analytic groups and lattices in $SL_2(\mathbb{C})$, Ann. Math. 118 (1983), 115–130.

[Ma24] Mannan, W. H. A fake Klein bottle with bubble, Bull. London Math. Soc. 56 (2024), 1605–1612.

[Ma68] Massey, W. S. Higher order linking numbers, J. Knot Theory Ramif. 7 (1998), 393–419.

[Ma69] Massey, W. S. Proof of a conjecture of Whitney, Pacific J. Math. 31 (1969), 143–156.

[Mc22] McDonald, C. Surface slices and homology spheres, arXiv: 2202.02696 [math.GT].

[Mi83] Milgram, R. J. Evaluating the Swan finiteness obstruction for finite groups, in *Algebraic and Geometric Topology, Proceedings, Rutgers 1983*, edited by A. Ranicki, N. Levitt and F. Quinn, Lecture Notes in Mathematics 1126, Springer-Verlag, Berlin (1985), 127–158.

[Mo73] Montesinos, J. M. Variedades de Seifert que son recubridores ciclicos ramificados de dos hojas, Bol. Soc. Mat. Mexicana 18 (1973), 1–32.

[Mo83] Montesinos, J. M. On twins in the 4-sphere I, Quarterly J. Math. 34 (1983), 171–199. II, *ibid.* 35 (1984), 73–83.

[My83] Myers, R. Homology cobordisms, link concordances and hyperbolic 3-manifolds, Trans. Amer. Math. Soc. 278 (1983), 271–289.

[Neu81] Neumann, W. D. A calculus for plumbing applied to the topology of complex surface singularities and degenerating complex curves, Trans. Amer. Math. Soc, 268 (1981), 299–344.

[NR78] Neumann, W. D. and Raymond, F. Seifert manifolds, plumbing, μ-invariant and orientation reversing maps, Lecture Notes in Mathematics 664, Springer-Verlag, Berlin (1978), 163–196.

[No69] Norman, R.A. Dehn's Lemma for certain 4-manifolds, Invent. Math. 7 (1969), 143–147.

[OS03] Ostváth, P. and Szábo, Z. Absolutely graded Floer homologies and intersection forms for four-manifolds with boundary, Adv. Math. 173 (2003), 179–261.

[Pr77] Price, T. M. Homeomorphisms of quaternion space and projective planes in 4-space, J. Austral. Math. Soc. 23 (1977), 112–128.

[Pu24] Putnam, A. The generalized Schoenflies theorem, www3.nd.edu/andyp/notes/Schoenflies.pdf.

[Qu01] Quinn, F. Dual decompositions of 4-manifolds, Trans. Amer. Math. Soc. 354 (2001), 1375–1392.

[Rok52] Rokhlin, V. A. New results in the theory of four-manifolds, Doklady Akad. Nauka SSSR 81 (1952), 221–224. [Russian]. French translation in [**GM**].

[Rol84] Rolfsen, D. Rational surgery calculus: extension of Kirby's theorem, Pacific J. Math. 110 (1984), 377–386.

[Ros84] Rosset, S. A vanishing theorem for Euler characteristics, Math. Z. 185 (1984), 211–215.

[Rub80] Rubinstein, J. H. Dehn's lemma and handle decompositions of some 4-manifolds, Pacific J. Math. 86 (1980), 565–569.

[Şa24] Şavk, O. A survey of the homology cobordism group, Bull. Amer. Math. Soc. 61 (2024), 119–157.

[Şa25] Şavk, O. Embeddings of homology spheres in sums of complex projective planes, in preparation.

[Sc83] Scott, G. P. The geometry of 3-manifolds, Bull. London Math. Soc. 15 (1983), 401–487.

[Se33] Seifert, H. Topologie dreidimensionaler Räume, Acta Math. 60 (1933), 147–238.

[Si05] Siebenmann, L. The Osgood-Schoenflies theorem revisited, Uspekhi Mat. Nauk. 60 (2005), 67–96. [Russian] Translation: Russian Math. Surveys 60 (2005), 645–672.

[St65] Stallings, J. Homology and central series of groups, J. Algebra 2 (1965), 170–181.

[Su75] Sullivan, D. On the intersection ring of compact three-manifolds, Topology 14 (1975), 275–277.

[Wd68] Waldhausen, F. On irreducible 3-manifolds which are sufficiently large, Ann. Math. 87 (1968), 56–88.

[Wa64] Wall, C. T. C. Quadratic forms on finite groups and related topics, Topology 2 (1964), 281–298.

[Wa65] Wall, C. T. C. All 3-manifolds embed in 5-space, Bull. Amer. Math. Soc. 71 (1965), 569–572.

[Wa66] Wall, C. T. C. Finiteness conditions for CW complexes II, Proc. Roy. Soc. Ser. A 295 (1966), 129–139.

[Wa67] Wall, C. T. C. Poincaré complexes: I, Ann. Math. 86 (1967), 213–245.

[Wa76] Wall, C. T. C. Classification of hermitian forms VI. Group rings, Ann. Math. 103 (1976), 1–80.

[Wi91] Wilson, J. S. Finite presentations of pro-p groups and discrete groups, Invent. Math. 105 (1991), 177–183.

[Ya97] Yamada, Y. Decomposition of S^4 as a twisted double of a certain manifold, Tokyo J. Math. 20 (1997), 23–33.

[Ze65] Zeeman, E. C. Twisting spun knots, Trans. Amer. Math. Soc. 115 (1965), 471–495.

Index

For EU product safety concerns, contact us at Calle de José Abascal, 56–1°, 28003 Madrid, Spain or eugpsr@cambridge.org.

www.ingramcontent.com/pod-product-compliance
Ingram Content Group UK Ltd.
Pitfield, Milton Keynes, MK11 3LW, UK
UKHW021914290526
471652UK00009B/429
* 9 7 8 1 0 0 9 7 1 5 3 8 6 *